Morphology and Infrageneric Relationships of the Genus *Jatropha* (Euphorbiaceae)

by Bijan Dehgan and Grady L. Webster

UNIVERSITY OF CALIFORNIA PRESS
Berkeley • Los Angeles • London

UNIVERSITY OF CALIFORNIA PUBLICATIONS IN BOTANY

Volume 74

UNIVERSITY OF CALIFORNIA PRESS
BERKELEY AND LOS ANGELES
CALIFORNIA

UNIVERSITY OF CALIFORNIA PRESS, LTD.
LONDON, ENGLAND

ISBN: 0-520-09585-5
LIBRARY OF CONGRESS CATALOG CARD NUMBER: 77-83116

Contents

Acknowledgments

Our sincerest thanks go to Professors John M. Tucker and Donald W. Kyhos for their careful reading and constructive criticism of this manuscript. And to Nancy Dehgan, without whose untiring effort, patience, and understanding, completion of this paper would not have been possible.

Financial assistance for much of the field work in Mexico was provided by the Chancellor's Patent Fund of the University of California at Davis and by a Sigma XI grant in 1975 to the senior author. Many of the plants used in this study were received from various institutions and botanical gardens. Most of the African species were donated by Mr. and Mrs. John Bleck of the Abbey Garden in Santa Barbara, California, to whom special thanks are herewith extended. Other sources include the University of California Botanical Garden at Berkeley; I.F.A.N.-Dakar, Republic of Senegal; the University of Ghana, Botany Department, Legon, Ghana; Botanical Garden, Bogor, Java; Desert Botanic Garden, Papago Park, Phoenix, Arizona; U.S.D.A. Federal Experiment Station, Mayaguez, Puerto Rico; Forest Research Institute, Dehra Dun, India; National Botanic Garden, Lucknow, India; Botanic Gardens, Georgetown, Republic of Guyana; Botany Department, Faurah Bay College, Freetown, Sierra Leone; Professor Otto Solbrig, Harvard University (seeds collected in Argentina); Department of Forests, Papua and New Guinea; Huntington Botanic Garden, Pasadena, California; and Instituto de Investigacao Agronomica de Angola.

Abstract

The genus *Jatropha* has been variously divided by many authors into distinct genera, subgenera, sections, and subsections. In the present study, past treatments of the genus have been re-evaluated primarily on the basis of anatomical, morphological, and cytological features. Two subgenera, 10 sections, and 10 subsections are recognized to accommodate the 165-175 known species. *Jatropha curcas* (section *Curcas*) is considered to be the most primitive member of the genus on the basis of (among other characters) its palmately lobed leaves, arborescent growth habit, and occasional hermaphroditic flowers. From it, or a similar ancestral taxon, evolution has proceeded toward specialization in vegetative structure, culminating in a facultatively annual growth habit in section *Jatropha* (subgenus *Jatropha*), and in rhizomatous shrub habit in section *Mozinna* (subgenus *Curcas*). These changes have been accompanied by reduction in reproductive structures in both subgenera. The evolutionary trends of the inflorescence in the genus have apparently followed two distinct pathways. On the one hand, modification and ramification of coflorescences and paracladia have resulted in the formation of highly symmetrical, compound dichasia in subgenus *Jatropha* (section *Peltatae*), or the coflorescence has been reduced to a single pistillate flower (as in section *Collenucia*). On the other hand, inflorescences have become drastically reduced to a few (or solitary) terminal or lateral flowers in subgenus *Curcas*. In both subgenera the position of the inflorescences has shifted from a lateral to a terminal position. The evolution of the flower, as in the inflorescences, has proceeded along two very distinct lines. In subgenus *Jatropha,* reduction and rearrangement of stamens has occurred with no concomitant change in the number of carpels, while in subgenus *Curcas* the number and arrangement of stamens has remained constant, but the carpels have become reduced from three to one. In spite of its widespread geographical distribution and much morphological diversity, *Jatropha* appears to be a natural genus as indicated by uniformity in a large number of characters, including basic chromosome number (n=11); typically crotonoid, aporate, inaperturate, binucleate pollen grains; brochidodromous leaf venation pattern; and (in most taxa) a typically dichasial inflorescence. A synopsis of infrageneric taxa in the genus with diagnostic keys and descriptions is provided, followed by a checklist of the species.

INTRODUCTION

This revision is an outgrowth of a hybridization study which began more than ten years ago. The study was prompted by observation of visual differences in floral and vegetative structures of the species within the genus *Jatropha* that seemed too profound for taxa of such presumed close affinity. Search of the literature revealed that, in spite of several taxonomic treatments and in spite of the fact that a number of species are cultivated for their ornamental and medicinal properties, *Jatropha* remains one of the least understood of all the genera in the Euphorbiaceae. This problem is compounded by the meager representation of the species in herbaria and by the poor condition of such material, due partly to the water-insoluble latex which oozes out during drying. Soon after this investigation began, it became apparent that any attempt to study the genus *Jatropha* must use living plants. This work is mainly based on wild and greenhouse-grown plants because it was the lack of such material that hampered earlier revisionary efforts by several taxonomists.

Jatropha is a morphologically diverse genus with a disjunct distribution in seasonally dry, tropical regions (Fig. 3); these intercontinental disjunctions appear to be explicable by the theory of continental drift (Dehgan, 1976). In the present study, the past treatments of the genus have been re-evaluated primarily on the basis of gross morphological attributes of reproductive structures. In interpreting distribution patterns of the taxa, a basic premise has assumed limited dispersability and subsequent geographical restriction of the great majority of the taxa. Of the two subgenera recognized, one, subgenus *Jatropha*, includes all African, Indian, South American, Antillean, and a few of the relict Meso-American species; the second, subgenus *Curcas* (with the exception of section *Curcas*), is confined to Mexico and adjacent regions, including extensions of the Sonoran and Chihuahuan deserts of Arizona and Texas. The affinities within and the relationships between subgenera, sections, and subsections that are recognized in the present publication have also been substantiated by various lines of evidence, especially from anatomical observations and experimental crosses.

There is considerable anatomical diversity within *Jatropha*. The number and arrangement of the vascular bundles in the petiole of various taxa range from 11, 9, 7, 5 in a ring, as free traces, medullated cylinder, or "U"-shaped free or medullated traces (Dehgan, in press). Laticifers also provide clues to the establishment of phylogenetic relationships (Dehgan and Craig, 1978). Three distinct types of laticifers can be distinguished: articulated, nonarticulated, and idioblastic laticiferous cells. The idioblastic laticiferous cells are characteristic of the most advanced taxa of subgenus *Curcas* in Mexico. "Chambered crystalliferous parenchyma" occurs only in species of subgenus *Curcas*, while other types of crystals are found scattered in the mesophyll of subgenus *Jatropha*. Morphology of the leaf epidermis, when studied with scanning electron microscopy, indicates the presence of "brachyparacytic" stomata in subgenus

Jatropha, and the true "paracytic" type in subgenus *Curcas.* Verrucose epidermal hairs are characteristic of subgenus *Curcas* (except section *Curcas*), whereas those of subgenus *Jatropha* are smooth. The more advanced taxa in the latter subgenus have uniseriate-multicellular hairs (Dehgan, 1976).

The results of interspecific hybridizations of 20 species in 8 sections support the suggested phylogenetic relationship of various taxa as presented in this paper (Fig. 4), and show a decrease in interspecific crossability parallel with evolutionary advancement (as seen in the morphological reduction series and drastic shifts in growth habit). Species of *Jatropha* are uniform in chromosome number with an auto-xenogamous breeding system, the floral mechanisms having moderate to pronounced interspecific differences. Related species are separated largely by genetic incongruity, and possibly by preferential fertilization, rather than by incompatibility. The phylogenetically more distantly related taxa are, for the most part, separated by actual genetic incompatibility barriers. Geographical isolation not withstanding, most species are capable of gene exchange within wide limits under experimental conditions (Dehgan, 1976).

The present work is not intended as a monographic treatment of the genus, but is rather a revision of the taxa above the species rank (subgenera, sections, and subsections). This revision is based on correlation of the gross morphological characters of the species with biosystematic and experimental data.

TAXONOMIC HISTORY

Jatropha appears to be a natural genus although it shows a great deal of morphological diversity. This diversity has been reflected in the disputes concerning its generic and subgeneric delimitations. Differing opinions have also been expressed regarding designation of a species as the generic type—a matter which the authors feel has not been satisfactorily resolved. The following chronologically arranged historical review, therefore, introduces some of the problems involved in the revision of the genus, and analyzes the question of generic nomenclature (including typification). It should be pointed out, however, that only pertinent Linnaean and post-Linnaean treatments are reviewed here and the reader is referred to the synopsis of taxa for specific references and appropriate floristic treatments.

Linnaeus (1737) first described the genus *Jatropha* as follows (Gen. Pl. 288):

Jatropha*. Manihot Tournef. 438. Dill. Elth. 173. Jussievia Houst. A. A.

*Masculini flores:

cal: perianthum vix manifestum.

cor: monopetala. Tubus brevissimus. Limbus quinque partitis laciniis subrotundis, concavis, patentibus.

stam: filamenta decem, subulata, in medio approximata, quinque alterna breviora. Antherae subrotundae, versatilis.

pist: rudimentum debile in fundo floris latet.

Feminini flores in eadem umbella cum masculinis:

cal: ut in masculinis.

cor: ut in masculinis.

pist: germen subrotundum, trisulcatum. Styli tres dichotomi. Stigmata obtusa.

per: capsula subrotunda, tricocca, trilocularis: loculis bivalvibus.

sem: solitaria, subrotunda.

In the contemporary *Hortus Cliffortianus* (1738), Linnaeus gave three diagnoses to which he later (1753) assigned binomial species names:

1. Jatropha foliis multipartitis laevibus, stipulis setaceis multifidis (*J. multifida*).
2. Jatropha foliis palmatis dentatis retrorsum aculeatis (*J. urens*).
3. Jatropha foliis cordatis angulatis (*J. curcas*).

3

Interestingly enough, in *Genera Plantarum* (p. 384) and in *Hortus Cliffortianus* he drew attention to the distinctiveness of *J. urens,* but as MacKenzie (1929) has pointed out, these peculiarities were not considered sufficient grounds by Linnaeus to justify recognition of a separate genus. Clearly species 1 and 3 are not in accord with Linnaeus' original description because they are both petaliferous.

In his *Species Plantarum* (pp. 1006-7), Linnaeus (1753) gave a full description of *J. gossypifolia* with a list of synonyms; also included were: *J. moluccana* [= *Aleurites moluccana* (L.) Willd.], *J. curcas* L., *J. multifida* L., *J. manihot* [=*Manihot esculenta* Crantz], *J. urens* [= *Cnidoscolus urens* (L.) Arth.], and *J. herbacea* [= *Cnidoscolus herbaceus* (L.) I. M. Johnston].

In the second edition of *Species Plantarum* (pp. 1428-1430), Linnaeus (1763) divided the above species into two groups with the first four species placed in "calyculati" and the last three listed under "acalyculati." It is clear that in spite of this final separation, his concept of generic delimitations was quite different from those of modern taxonomists.

Adanson (1763), in his *Familles des Plantes,* recognized two genera which were apparently based on Linnaeus' *Species Plantarum,* as indicated below:

	Feuilles	Fleurs	Calice.	Corolle.	Stam.	Stigm.	Fruit.	Graines
Curcas, Clus. mandubi. guacu. Marg. 97 Jatropha 3 Lin. sp.	alt.	Corymb. axil.	6 feuill.	5 pet.	10	6 cilind.	Capsule à 3 log. 3 valv.	2 dans ch. log. ovoid.
Jatropha. 6 Lin. sp. Jussievia. Houst. ic. 3 Manihot. Dill. Elt. s. 173	Id.	Corymb. term.	Mal. tub. long. 5 div. Fem. t. court 5 dents.	Mal. 0 Fem. 5 pet.	10	15 à 30 cil.	Id. 6 valv.	Id. spher.

The numbers quoted by Adanson refer to the numbers assigned to the species of *Jatropha* by Linnaeus in 1753. Thus the type species of the genus *Curcas* Adans. is, by implication, *Jatropha curcas* L.; this, incidentally, makes it impossible to designate *J. curcas* the lectotype of *Jatropha,* as was done by Hitchcock and Green (1935). Adanson was the first to remove the cassava plant *Manihot* as a separate genus; although the name *Manihot* was first validly published by Miller (1754), that author used it as a substitute for Linnaeus' *Jatropha.* This leaves two elements in *Jatropha* sensu Adanson: *Jatropha* s. *str.* of current authors and *Cnidoscolus.* Despite Adanson's failure to distinguish between *Jatropha* and *Cnidoscolus,* his work represents the first advance in the clarification of generic concepts and is far superior to the circumscriptions of most other eighteenth century authors such as Jussieu (1789), who still combined *Jatropha* with *Manihot.*

The definitive realignment of generic limits was achieved by Pohl (1827), who recognized four genera: *Adenoropium, Jatropha, Manihot,* and *Cnidoscolus.* Pohl's *Adenoropium* was a renaming of *Jatropha* auct., since he restricted the name *Jatropha* to *J. curcas;* this was of course illegitimate, since Adanson had already placed *J. curcas* in a segregate genus *Curcas;*

but although the nomenclature of the two authors differed, their concepts were basically the same. The particular merit of Pohl's disposition was the recognition of *Cnidoscolus* as a distinct genus and, in view of the well-documented nature of his treatment, it is rather difficult to understand why later authors did not follow him.

Baillon's (1858) genus *Jatropha* is apparently equivalent to Pohl's *Adenoropium.* The two species *J. curcas* L. and *J. hernandiifolia* Vent. were included in Adanson's genus *Curcas.* Distinction of the two genera was apparently based on the presumably gamopetalous nature of the corolla in *Curcas* ("Les Curcas ont la corolle gamopétale").

Grisebach (1859) recognized Baillon's taxa as sections of *Jatropha.*

> Sect. I. Adenorhopium*–petals distinct or cohering at the base, spreading.
> Sect. II. Curcas–corolla sympetalous, styles cohering below.

Cnidoscolus was considered distinct from *Jatropha.*

The first truly comprehensive treatment of the genus was presented by Jean Mueller, initially in De Candolle's *Prodromus* (1866) and later in Martius' *Flora Brasiliensis* (1874). Three sections were described with three subsections in the first section, and six divisions without specific taxonomic rank in the second; *Cnidoscolus* constituted the third section.

> Conspectus sectionum:
>> Sect. I. Curcas Griseb. Petala basi tenaciter imbricatum conglutinato cohaerentia.
>>> Subsect. I. Loureira–Genus Loureira Cav.
>>> Subsect. II. Eucurcas Muell. Arg.
>>> Subsect. III. Mozinna–Genus Mozinna Ortega.
>> Sect. II. Adenorhopium Griseb., Genus Adenorhopium Pohl.

The majority of the species enumerated were described by Mueller himself. The inclusion of *Cnidoscolus* within *Jatropha* was justified by Mueller on the basis of the theoretically suppressed (as opposed to truly absent) petals of the former—an ingenious but untenable assumption.

A comprehensive monographic treatment of the genus was attempted by Pax (1910) in Engler's *Das Pflanzenreich.* In addition to classification of all the species known to him, diagrams of phylogenetic affinities (Fig. 1), and a detailed discussion of the geographical distribution of the genus, were presented. Unfortunately, the genus *Cnidoscolus* was once again included as a subgenus of *Jatropha,* a decision perhaps influenced by Mueller's treatment (1866). It was, however, considered to be only distantly related to the subgenera *Adenorhopium* and *Curcas.* The fact is quite apparent in the phylogenetic diagram (Fig. 1), and also in the later works of Pax and Hoffmann (1919, 1931) where *Cnidoscolus* was treated as a distinct genus. The criterion here, as elsewhere, is merely the presence of petals in *Jatropha* and their absence in *Cnidoscolus.*

As indicated in Figure 1, two subgenera (excluding *Cnidoscolus*), eight sections, and twelve subsections were recognized. The two subgenera were separated on the basis of a single character, namely the coherence or distinctness of the petals. Hence:

> Petala libera vel basi tantum vel vix cohaerentia.
>> Subgen. I. Adenorhopium (Pohl) Griseb.
> Petala ± cohaerentia.　　Subgen. II. Curcas (Adans.) Griseb.

*Spelling used by Grisebach.

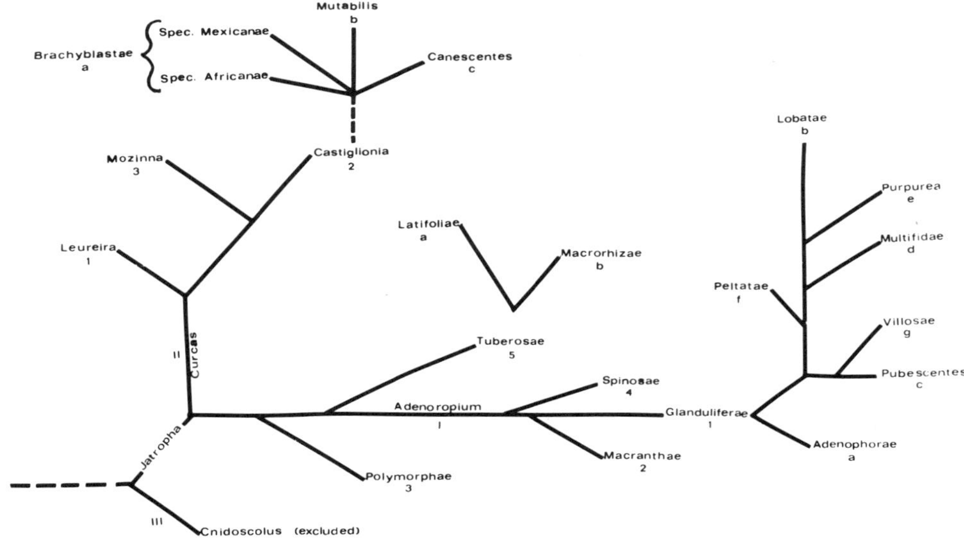

Figure 1. Phylogenetic affinities of subgenera, sections and subsections of *Jatropha* (reassembled from Pax, 1910).

McVaugh (1945b) discarded these two subgenera on the grounds that "arrangement of the genus based on a single corolla character is actually too simple to be a natural one." While we are in complete agreement with McVaugh's objection, the separation of the two subgenera was clearly based on the same criterion that had been used by other authors and was not original with Pax. The foliar characters (length of petiole, presence or absence of glands, and degree of lobing of the leaf) which form the basis of separation of Pax's sections are so variable and inconsistent that they are useful only in recognition of species. While Pax's treatment is a monumental contribution to understanding of the genus *Jatropha,* its value lies more in its utility for identification than in its evolutionary model.

Cnidoscolus was treated as a separate genus by McVaugh (1944) on the basis of its white single floral envelope, stinging hairs, and distinctive petiolar glands. He cited numerous other morphological and anatomical characters, and pointed out that perhaps *Jatropha* from an evolutionary standpoint is closer to *Manihot, Aleurites,* and even *Hevea* than it is to *Cnidoscolus.* This belief has more recently been firmly established by Miller and Webster (1962) on the basis of pollen morphology, number of vascular bundles in the petiole, and difference in chromosome numbers (2n = 22 in *Jatropha* and 2n = 36 in *Cnidoscolus*).

McVaugh's treatment of the American species of *Jatropha* (1945b) attempted to put forth the principal lines of divergence in the genus. Four sections were proposed which were presumably more "homogeneous" than the sections accepted by previous authors. Indeed, if one considers the gross floral and vegetative morphology of the species in each section (with a number of exceptions), one is overwhelmed by similarities within and dissimilarities among the first three sections (*Polymorphae, Macranthae,* and *Adenoropium*). The fourth section, *Mozinna,* is clearly unnatural, and the inclusion of the species within that section, as pointed out by McVaugh himself (l.c.), was influenced by the fact that they are for the most part Mexican. A brief discussion of these four sections as circumscribed by McVaugh follows:

1. Section *Polymorphae. J. integerrima* Jacq., *J. angustifolia* Griseb., *J. tupifolia* Griseb., *J. paxii* Croiz., and *J. minor* Urb. are included in this section. *Jatropha hastata* Jacq., *J. acuminata* Desr., *J. pandurifolia* Andr., *J. diversifolia* A. Rich., *J. moluensis* Sessé & Moc., *J. glaucovirens* Pax & Hoffm., and *J. pauciflora* Griseb. are considered to be synonyms of *J. integerrima* Jacq. Furthermore, the five African-Arabian species which are attributed to this section by Pax (1910), including *J. unicostata* Balf.f., *J. variegata* (Forsk.) Vahl, *J. capensis* (L.f.) Sond., *J. variifolia* Pax, and *J. prunifolia* Pax, are said to be distantly related. The West Indian aggregation of species in McVaugh's circumscription represent a natural group. The addition of several other taxa in the present revision, however, will result in a much more widespread distribution within the American tropics.

2. Section *Macranthae.* The principal leaf characters upon which Pax's subsections were based (degree of lobing, amount of pubescence, etc.) were found to be inadequate for disposition of species by McVaugh, who relied on floral characters. Indeed, this section is sufficiently homogeneous morphologically that it has been kept intact in all earlier treatments of the genus including that of Mueller (1865), who placed the species (with but very few exceptions) in the six divisions of his section *Adenoropium.* Three of the species included in *Macranthae* should be noted. *Jatropha mutabilis* (Pohl) Baillon, placed in section *Mozinna* by Pax, was transferred. *Jatropha macrantha* Muell. Arg. and *J. macrocarpa* Griseb. were considered to be aberrant species for this section, although *Jatropha macrantha* was designated as the type species. As will be noted later, however, *J. macrantha* and *J. macrocarpa* are more properly included in section *Polymorphae.* The removal of the type species has necessitated the change of the section name from *Macranthae* to *Peltatae.*

3. Section *Adenoropium* (section *Jatropha* of the present paper). McVaugh recognized four unnamed groups here, of which one has 10 stamens and the other three groups, 8 stamens. *Jatropha macrorhiza* Benth., which was placed in this section, has apparently been mistakenly diagnosed—it has 8 stamens, not 10, and lacks the tubular corolla. As will be noted later in this paper, the species actually has closer affinity to section *Polymorphae* than to *Adenoropium.* Section *Adenoropium* is otherwise quite homogeneous and, as pointed out by McVaugh, includes a number of African and Indian taxa as well as American. One would also have to agree with McVaugh that most species in the *J. gossypiifolia* complex are so closely related that their separation is possible only with extreme difficulty. In this connection it should be noted that for the most part no keys were provided for separation of the species within the groups.

4. Section *Mozinna.* Of the two subsections recognized by McVaugh in this section, one, *Eucurcas* (p. 283), is reasonably homogeneous. Subsection *Mozinna,* as admitted by him, is the least natural: "the characters used by Mueller and Pax to divide the group seem to produce an unnatural arrangement and I am at a loss to effect any division within the subsection unless *J. gaumeri* Greenm., an endemic of the Yucatan Peninsula, be removed entirely from the section *Mozinna* on the basis of its unequally divided style." Excluding *J. hernandiifolia* Vent. and *J. divaricata* Sw., which are recognized as a subsection of *Polymorphae* in the present revision, the species included by McVaugh in subsection *Mozinna* have complicated reticulate evolutionary relationships which require experimental study (see Dehgan, 1976).

Wilbur (1954) added two new species (*J. bartlettii* and *J. hintonii*) to the existing three in the subsection *Eucurcas* of the section *Mozinna* of McVaugh, bringing the total of the American species to five. Although it is difficult to determine the origin of *J. curcas,* Wilbur points out that it is probably Mexican, because "it was without doubt part of the flora of Mexico and probably of northern Central America before the arrival of Cortez, and it most likely originated there . . . the subsection, hence, appears to be one which originally was nearly or completely restricted to Mexico." This statement is questionable because the subsection *Eucurcas* (*sensu* Pax, 1910) also includes two east African and one Indian species.

THE PROBLEM OF TYPIFICATION OF *JATROPHA*

Since differing opinions have been expressed regarding typification of the genus *Jatropha* and several references have been reviewed above that bear directly on this problem, a brief discussion here seems appropriate.

McVaugh (1944) rejected the proposal by MacKenzie (1929) that *Jatropha manihot* L. be considered the typical element. If MacKenzie's judgement were accepted, it would mean a complete reversal of botanical usage, because the genus *Manihot* (Cassava) has been accepted as distinct from *Jatropha* (Physicnut) for more than 200 years. Among modern authors, only Small (1933) has adopted such a concept. *Jatropha curcas* L. was also rejected as the type of the genus by McVaugh (l.c.) on the grounds that this species was already the type of Adanson's segregate genus *Curcas,* although *J. curcas* has been cited as the type by some recent authors (Hitchcock and Green, 1935; Lourteig and O'Donnell, 1943). Because of the detailed description given for *J. gossypiifolia* L. in the first edition of the *Species Plantarum* (Linnaeus, 1753), McVaugh proposed to designate that species as the generic type.

We do not propose discarding the two names that are well established in botanical literature by replacing the name *Jatropha* with *Manihot* and, in turn, *Manihot* with *Janipha* (a classically 'correct' name). Thus, any extended discussion here would not be appropriate. However, it should be pointed out that McVaugh's choice of *J. gossypiifolia* as lectotype is rather questionable. Adanson (1763), in his citations for *Jatropha,* listed *J. multifida* L. twice (as *Species Plantarum* no. 6 and as the "Manihot" of Dillenius) and did not mention *J. gossypiifolia.* McVaugh has argued for the choice of *J. gossypiifolia* as lectotype because of the extended description of this species in the *Species Plantarum* (1753), based on van Royen. However, the confused description by Linnaeus in the *Genera Plantarum* (1754), although obviously based in part on the details given for *J. gossypiifolia,* also includes features (missing "calyx," hypocrateriform "corolla") which apply to either *Cnidoscolus* or *Manihot.* In our opinion, the citations of Adanson come close to lectotypifying *Jatropha* by *J. multifida* L., especially on the basis of his citations of the plate in the *Hortus Elthamensis* of Dillenius (1732). Nevertheless, since Adanson also included *Cnidoscolus* in his concept of *Jatropha,* it can be argued that he did not achieve a binding lectotypification. To adopt *J. multifida* as lectotype now would require renaming many of the subgeneric taxa used by Pax and by McVaugh, without achieving any greater clarification of concepts. Furthermore, there seems to us no likelihood whatever that *J. gossypiifolia* and *J. multifida* will ever be placed in different genera. Consequently, we somewhat reluctantly concur with McVaugh's choice of *J. gossypiifolia* as lectotype on the grounds of tradition and expediency.

MORPHOLOGY

The remarks that follow are primarily based on observations of living specimens, which have been grown in the greenhouse for the past several years, and on notes made in the field in Mexico.

1. HABIT OF GROWTH

Jatropha has been variously described as comprising trees, shrubs, and perennial herbs (Mueller, 1866, 1874; Pax, 1910; Pax and Hoffmann, 1919, 1931; Hutchinson, 1913; McVaugh, 1945b; Ingram, 1957; Webster, 1967, 1968; Standley and Steyermark, 1949; et al.). The variation within the genus is so great that, with the exception of hemicryptophytes, vines, epiphytes, and parasites, all other classes suggested by Raunkaier (1934) and reviewed by Lacza and Fekete (1969, 1972) and Fekete and Lacza (1970, 1971) could be used to describe the various growth forms and growth habits.

A. Growth Forms

1. Therophytes. Although no truly annual species exist in the genus, one species, *J. gossypiifolia,* approaches this condition. This point has been noted by Standley and Steyermark (1940), who described this plant as "probably annual." We have observed that *J. gossypiifolia* is the only species among the many grown in the greenhouse which cannot be carried from one year to another except with extreme difficulty. Despite some secondary growth in the stem, the plant either dies out completely or remains weak and does not resume normal growth in the second season. This is the most widespread species in the genus, and is commonly found as a weed in vacant lots and around cultivated fields in the Old and New World tropics. It can attain a height of up to 2 m in a single growing season and complete two consecutive generations.

Jatropha gossypiifolia, therefore, can be described as a facultatively annual, herbaceous subshrub with woody stem. The growth is sympodial, and apical dominance is apparently absent because the plants branch while quite young and long before flowering. This description, however, does not completely apply to closely related species (e.g. *J. excisa*), since they are definitely perennials.

2. Geophytes. These are perennial plants with their perennating organs (viz. tubers) underground and protected; stems die above ground. Only two North American-Mexican species, *J. macrorhiza* (Pl. I, figs. C and D) and *J. cathartica,* and apparently two South American species, *J. isabellii* and *J. dissecta,* have this characteristic. Several South and tropical African spe-

cies may also be included here, e.g. *J. gallabatensis, J. zeyheri, J. woodii* (Pl. II, fig. A), etc. With the notable exception of *J. cathartica,* all were placed in section *Tuberosae* by Pax (1910). Plants in this group are sub-shrubs with herbaceous shoots and with a woody-succulent, more or less subterranean caudex (Fernald, 1950) from which one to several shoots may appear annually during the growing season. These shoots elongate quite rapidly until the onset of flowering, usually within a month. At this time, apical dominance is apparently removed and a few much shorter laterals appear which soon cease growth by ending in an inflorescence.

3. Phanerophytes. A majority of the species of *Jatropha* fall into this class with the following five sub-classes recognizable:

a) Erect fleshy sub-shrubs with woody-succulent, above-ground caudex and branches. This group is evidently best represented in Africa; *Jatropha podagrica* represents the only American species (Pl. I, fig. B). If the term is applied broadly, however, certain subspecies (races, *fide* McVaugh, 1945b) of *J. cinerea* and *J. cuneata,* as well as several other Central and South American species, may also be included here. The African species are exemplified by *J. marginata, J. paradoxa* (Pl. II, figs. C and D), *J. capensis* (Pl. II, fig. B), *J. velutina* (Pl. II, fig. E), etc. The branching in this group is somewhat similar to that of geophytes except that apical dominance seems to be absent. Following germination, the leader continues growth while the caudex is enlarging. Lateral branches appear later at equidistant intervals on the upper portion of the caudex. These secondaries usually remain considerably shorter than the main axis. Flowering does not alter this pattern of growth. The apical meristem of the branches is pushed aside by the inflorescence and will resume normal growth (see below), in spite of the seemingly terminal inflorescence.

b) Decumbent woody sub-shrubs with ± swollen base and succulent branches. Included here are *J. giffordiana,* certain races of *J. cinerea, J. cuneata* (Pl. IV, fig. E), and *J. macrantha* (Pl. I, fig. E). These species, for the most part, retain this characteristic under cultivation. The plants have a maximum height of 1-3 feet under natural conditions and an extremely deep root system (cf. Pl. IV, fig. E), probably because they occur in sandy areas of very low rainfall.

c) Erect shrubs arising from subterranean rhizomes, at times forming pure stands of many branches covering several feet. Examples of this type, as far as the authors know, include only three species: *J. dioica* (Pl. IV, figs. A and C), including the two varieties *sessiliflora* and *graminea* (McVaugh 1945a); *J . cardiophylla* (Pl. IV, fig. B); and a third, closely related species (or possibly a variety of *J. cordata*) which is found in dry, south-facing slopes between Guadalajara and Zacatecas in Mexico. It should be noted that the type of *J. cordata* has been described as a small tree, and field observations in a number of localities in the Sonoran Desert indicate no rhizomes (cf. Pl. III, fig. A). In this connection, an apparent form of *J. cardiophylla,* which is upright and nonrhizomatous, was recently collected at Alamos, Mexico, by J. Dodson of the University of California at Berkeley. This form has not been observed by the authors either in Mexico or in Arizona. The flowering habit also differs slightly in this form by being a typical Jatrophoid dichasium rather than the loosely branched or sessile, few-flowered inflorescence which is found in the rhizomatous form.

The rhizomatous character of the three species reported here has hitherto been overlooked in previous accounts, including those of the original descriptions, but has been referred to as "root crown" by Cannon (1911) and by Shreve and Wiggins (1964). As indicated earlier, the "rhizomatous" species reported in the African flora (Hutchinson, 1913), e.g. *J. gallabatensis,* are similar to *J. macrorhiza* and *J. cathartica,* and are quite distinct from the three species above in having subterranean, tuberlike structures rather than rhizomes.

d) Erect shrubs 1-3 m tall, usually multibranched from near the base. With the notable exception of *J. hernandiifolia* and *J. integerrima* (Pl. I, fig. F), both of which are West Indian and show no succulence whatsoever, this group includes many of the Old and New World species which may be characterized as shrubs or sub-shrubs with succulent stems and branches, e.g. *J. cuneata* (Pl. IV, figs. E and F). The growth is sympodial, although a main leader may grow for a time, following seed germination, but lateral branches appear soon after as axillary buds start to expand. "Leader displacement" (Radford et al., 1975) does take place, however, as the shoot meristem is pushed aside by the inflorescences. From two to several successive laterals take over the leader, resulting in what may be called secondary leader displacement, but usually do not result in the formation of an "umbrella shaped" plant (Radford et al., 1975).

e) Large shrubs or small trees reaching 10 m or more and usually with thickened trunk (resembling an elongated caudex) and branches. This group is distinguished from the preceding by height and by lack of branches on the lower portion of the plant, which, by definition, would constitute trees. Some difficulty may arise, however, in such species as *J. augustii* and *J. hieronymii,* because lower branches do arise occasionally from near the base, thereby making them large shrubs rather than trees. Other species include *J. curcas* (Pl. III, fig. B), *J. multifida* (Pl. I, fig. A), *J. malacophylla* (*sensu* McVaugh 1945b), *J. platyphylla, J. gaumeri,* and *J. cordata* (Pl. III, fig. A), as well as a number of other Central and South American species. The Old World species of *Jatropha* in this category are few: *J. glandulifera* from India, which has been described by Cooke (1906) to be a small evergreen tree, and taxa which are related to *J. curcas,* e.g. *J. afrocurcas, J. macrophylla,* and *J. villosa.*

According to Hallé (1971), *J. multifida* shows a segmented, but linear architecture (no distinction is made between this term and growth habit—see Tomlinson and Gill, 1973 for discussion) as a consequence of differentiation of the inflorescence which results in the apical meristem being pushed aside. Although this condition is present in most of the species in section *Peltatae* (*sensu* Dehgan and Webster) and in *J. platyphylla,* the remainder of the species mentioned above show leader displacement by two or three equal laterals, resulting in a segmented but three-dimensional architecture as in *Manihot esculenta* (Hallé, 1971). This growth pattern is referred to as "dichotomous substitution" (Radford et al., 1975) or "isotomous branching" (Foster and Gifford, 1974).

Jatropha augustii and *J. hieronymii* can neither be considered sympodial in the strictest sense, nor monopodial. There exists a main axis which continues growth by lateral displacement of the apical meristem as a result of inflorescence differentiation, as in *J. multifida.* However, at irregular intervals lateral branches arise that are nearly equal in strength and height to the main axis and without regard to the presence or absence of an inflorescence and irrespective of age. This condition may be referred to as a pseudosympodial type of branching or perhaps "dichasial sympodium" as designated by Troll (1935).

B. Shoot Dimorphism

The subject of shoot dimorphism in *Jatropha* has received little attention by taxonomists or morphologists studying subtropical desert plants. Short or arrested shoots are characteristic of at least six Mexican-North American species; these include *J. dioica, J. cardiophylla, J.*

cuneata, J. neopauciflora, and the *J. cinerea* complex. The extreme case of arrested shoots is shown in *J. canescens,* where only swollen areas resembling nodes occur on the main axis at regular intervals. From those nodes short shoots of about 3-4 cm terminating in an inflorescence appear in the axils of the leaves. This situation is analogous to the one described for *Ginkgo biloba* by Foster (1938). The shoot apices in *Jatropha,* as in *Ginkgo,* are reversible, and short shoots may be induced to become long shoots by removing the apical portion of the plant. This fact, in addition to the reduction in number of such shoots under greenhouse conditions, may indicate a complex physioecological relationship where production of auxin, which causes apical dominance (Sachs, 1965; Gifford and Corson, 1971), is directly linked to the length of the growing season and amount of available moisture. Both of the latter are greatly increased in the greenhouse as compared to natural desert conditions. Further elaboration on this subject is beyond the scope of this revision and clearly requires experimental evidence.

The condition described here is not restricted to *J. canescens;* it occurs in all the species mentioned above. Several species of *Jatropha* from the Somalia-Arabia regions have been reported by Hutchinson (1913) to have short shoots similar to the American species, e.g. *J. rivae, J. robecchii, J. aspleniifolia, J. spinosa* var. *somalensis, J. ferox, J. crinita,* etc. This clearly independent origin of short shoots in two such widely separated geographical regions may indicate parallel evolution, a hypothesis which is also supported by the striking similarity of floral structure (see below).

C. Methods of Extension Growth

The significance of this subject was recently brought to light by Tomlinson and Gill (1973). They recognized three distinct types of extension growth, primarily based on the presence or absence of periodicity of growth. The richness of architecture in Euphorbiaceae described by Hallé (1971) can perhaps be shown in *Jatropha* alone, which can offer examples in all three categories:

1. Articulated growth. Plants in this group are characterized by annual increments of shoot extension with a distinct, morphological discontinuity or articulation between each increment. The majority of *Jatropha* species that occur in the drier parts of South and Central America show this characteristic even under greenhouse conditions. Dormancy in such species is apparently related to the frequency of rainfall and to temperature-light fluctuations. If such plants are placed under long days and watered when dry, they continue growth without any sign of dormancy. Conversely, dormancy can be induced by merely withholding irrigation or reducing day length. It should be pointed out that continuous growth under long days in these species results in an imbalance of pistillate or staminate flower production. Examples in this group are too numerous to enumerate, but include such familiar species as *J. curcas, J. multifida, J. cinerea,* and *J. cordata.*

2. Non-articulated growth. No dormancy, and consequently no distinct articulations separated by bud scales, exists in these species. This condition has been referred to by Koriba (1958) and Tomlinson and Gill (1973) as "evergrowing." Two species, *J. integerrima* and *J. hernandiifolia,* may be given as examples. Ecological conditions under which these and possibly other West Indian species grow indicate continuous equable temperature and a more or less even distribution of rainfall throughout the year (Seifriz, 1943; Howard, 1973; and see Sarmiento, 1976 for variation in plant communities). Ecological conditions combined with the complete absence of apical dominance result in continuous, uninterrupted growth and flowering.

3. Intermediate growth. This has been defined as articulated growth without complete dominance, as in *J. podagrica* and *J. capensis.* There is no complete dormancy in these species. They continue to grow and flower during the winter months, but very slowly and in sudden flashes which produce distinct but relatively obscure bud-scale scars. The leaves at this stage are quite small and concentrated at the tip of the branches. A similar condition has been observed in the Sonoran Desert, where during the dry season a few small leaves exist at the tip of the branches in several species including *J. malacophylla* and *J. platyphylla.* The dormancy in these species, however, becomes eventually complete and the plant becomes leafless.

There is little doubt that such species as *J. cinerea* (*sensu lato*) present a special problem and require further study. A great deal of variation is observed not only in their growth habit and gross morphology, but also in the state of dormancy; some plants are quite leafless during the dry season, while others in the same locality are green, growing, and even occasionally producing flowers. The same situation also exists under greenhouse conditions in spite of uniform circumstances. No satisfactory factual or hypothetical explanation for this phenomenon can be offered at this time, other than perhaps genetic differences between individuals.

2. ROOTS AND TUBERS

Detailed anatomical studies of the root could conceivably provide taxonomic evidence for the elucidation of the categories in *Jatropha.* But several years of greenhouse and field observation of their gross morphology have proved to the contrary, and it seems unlikely that additional documentation would show much significance. Certain generalities, however, merit consideration.

The five roots that are mentioned later in connection with seedlings are organized in such a way that a central and four outer roots are present in all species so far studied. The central root is persistent, well-developed, and invariably becomes fleshy and grows quite deep, forming the primary tap root. The four outer roots are organized equidistantly from each other and from the central one. These form the more shallow roots that in xerophytic species frequently appear as succulent roots above ground (e.g. *J. cinerea*—Pl. III, fig. E). The basic arrangement does not change in species of more mesic habitats, but the outer roots grow more or less horizontally and near the soil surface (e.g. in West Indian and most of the South American species).

In species with subterranean 'tubers' (e.g. *J. macrorhiza*—Pl. I, fig. D and *J. woodii*—Pl. II, fig. A) the hypocotyl is very short and the area immediately below it becomes thickened after germination to form the so-called tuber. The distinction between the root and the tuberous portion is, therefore, extremely difficult to make.

Anatomically the growth is anomalous and vascular bundles occur at irregular intervals. The pith is present and small relative to the diameter of the 'tuber,' but no transitional region is evident. Lateral roots which appear at the seedling state on this swollen area leave prominent scars on the tuber when mature (Pl. I, fig. D). It seems more plausible, therefore, that the swollen portion be considered the 'tuber' on which adventitious roots arise. This pattern is analogous to the development of fleshy roots in *Daucus carota* (Esau, 1940) as a result of massive development of parenchyma in the phloem and xylem of the root.

Cannon (1911) referred to rhizomes of *J. cardiophylla* as "horizontal extensions of the root system." Little doubt exists on the rhizomatous structure of these organs, along which roots appear in concentrated areas (Pl. IV, fig. B). Taxonomically this is significant, since as far as is known, this condition exists in only two other species in addition to *J. cardiophylla,* namely *J. dioica* and an undescribed species closely related to *J. cordata.* Of the three species, the first two belong to section *Mozinna* and the third to section *Loureira,* both of which are in subgenus *Curcas.*

Tuberlike storage organs occur in *J. capensis* (Pl. II, fig. B) and may indeed be unusual for the genus. These storage structures do not produce shoots and are merely swollen portions of the root made up of parenchymatous tissue containing many starch grains (as evidenced by the IKI test).

3. LEAVES

In contrast to the uncomplicated leaves of other subfamilies of Euphorbiaceae (e.g. Phyllanthoideae, Webster 1956, 1958), those of Crotonoideae in general, and the genus *Jatropha* in particular, show a great deal of diversity. Several morphological and anatomical features have proven to be useful in the delimitation of sections and subsections and are discussed in other publications (Dehgan, 1976; Dehgan and Craig, 1978). This, therefore, will discuss generalities rather than details of anatomical and morphological variations.

A. Arrangement

Without exception, the arrangement of leaves in the genus is alternate with spiral phyllotaxis. The seemingly whorled leaves that are occasionally observed at the tips of actively growing, normal shoots and the "opposite" leaves on spur shoots are also alternate when examined closely; deviations in leaf arrangements are otherwise not known to occur.

B. Lamina

The size of the leaf may vary from less than 1 cm (e.g. 2-3 mm in *J. nogalensis* and *J. arguta*) in the extreme xeric habitats of the Somalian deserts (Chiovenda, 1929) to the largest known leaf in *J. platyphylla* which measures 35-40 cm long and wide. Two generalizations seem to hold with regard to leaf size:

1. Plants occurring under more xeric conditions on both the African and American continents have the smallest leaves, whereas species that grow in more mesic habitats possess the largest leaves, as expected on ecological grounds. However, this excludes such species as *J. cathartica, J. macrorhiza,* and the like, which have subterranean storage organs and are, therefore, not subject to the same environmental stresses as other taxa (e.g. *J. dioica, J. cuneata,* etc.).

2. African species in general seem to have considerably smaller leaves than those of American species. South American species (e.g. in section *Peltatae*) generally have larger leaves than those of meso-America.

Leaf architecture in *Jatropha* often provides an important diagnostic feature in the delimitation of species. An exception to this is found in section *Jatropha,* where extreme similarity of leaf form has resulted in poor treatment of species by Pax (1910) and their referral to species groups by McVaugh (1945b). According to McVaugh (1945b) the more than thirty American species reported in this section could be condensed into about four "by a not-too-conservative systematist." One would have to sympathize with McVaugh, for the American taxa differ mainly in depth of the lobes and angle of the sinus between them. The African species belonging to this section (subsection *Pubescentes*), on the other hand, are sufficiently differentiated to pose no difficulty.

Section *Jatropha* excepted, all the other sections are so variable with regard to leaf morphology that it is difficult to write general descriptions for them. For example, leaves in section *Polymorphae* Pax (Pl. V), as the name implies, display shapes that vary from cordiform (*J. integerrima*–Pl. V, fig. E) to hastiform (*J. tupifolia*–Pl. V, fig. D) or narrowly oblong-lanceolate (*J. angustifolia*–Pl. V, fig. G), and exhibit various degrees of lobing (*J. macrorhiza*–Pl. V, fig. C; *J. macrantha*, etc.). The leaves in section *Peltatae*, on the other hand, are characteristically deeply or shallowly lobed; lobes may vary from 3 in *J. podagrica* to 11 or more lobes or divisions in *J. multifida*, but may be peltate and ciliate-glandular as in *J. augustii* (Pl. VI, fig. D).

Subgenus *Curcas* presents a highly heterogeneous appearance as far as leaf shape is concerned. Excluding section *Curcas*, which has 5-7 shallowly lobed leaves (Pl. VII, figs. A and B), the leaves in the remainder of the taxa in the subgenus can be cordiform (*J. cordata*–Pl. VIII, fig. C; *J. vernicosa*), hastiform (e.g. certain 'races' of *J. cinerea*), reniform (*J. canescens, J. cardiophylla*–Pl. IX, fig. G), spathulate (*J. neopauciflora, J. sympetala*–Pl. IX, figs. B and C, and short shoot leaves of *J. dioica* and *J. cuneata*–Pl. IX, figs. D and F), or variously lobed (*J. dioica, J. cuneata*–Pl. IX, figs. D-F; *J. ciliata*–Pl. VII, fig. D), etc. Noteworthy is the fact that *J. cinerea* (*sensu* McVaugh) encompasses all the leaf types mentioned here in addition to being 3-5 lobed and having many intermediate forms. This species, as well as *J. integerrima* (*sensu* McVaugh 1945b), should be considered as representing complexes of species which probably because, of introgressive hybridization, have for the most part lost their original integrity. McVaugh (1945b) recognized this problem to some extent and alloted a brief discussion to both species, although he does not mention introgressive hybridization as a cause for such diversity, but merely designated "geographical races" in *J. cinerea* and relegated at least four species to synonymy under *J. integerrima* Jacq.

Considerable variation is also observed in the basal and apical portions of the leaves. The cordate condition seems to be predominant in the genus, although cuneate, hastate, reniform, and more or less sagittate bases also occur. The leaf apices, on the other hand, range from acute-acuminate to rounded. The margins may be entire, aculeate (usually with mucronate teeth), ciliate (usually glandular), serrate or doubly serrate (with or without glands), or may be sinuate or undulate, but never lacerate or divided unless infected by virus (such plants are frequently encountered under natural conditions).

Variability in the leaf texture in the genus is considerable and, although no truly scabrid surface occurs, the leaves are more or less hyaline and may be firmly membranous (e.g. *J. gaumeri, J. gossypiifolia*, etc.) or coriaceous (as in *J. podagrica, J. augustii*–section *Peltatae* in general) becoming somewhat fleshy (carnose) in some African (*J. velutina*) and some American taxa (*J. cinerea*). The vestiture may range from completely glabrous to densely tomentose or villous.

C. Venation

In contrast to the extreme variation of the leaf shape, the venation is uniformly "Brochidodromous-Semicraspedodromous": secondaries are joined together in a series of prominent arches, but one of the two divisions continuing to the margin ends blindly, or at the glands or serrations when present (cf. Hickey 1973; Dilcher 1974; Sehgal and Paliwal 1974; and see Pls. V through IX). The only exception occurs in *J. augustii* where no arches are present and the "actinodromous" veins continue to the margins directly (Pl. VI, fig. D). Although the vein endings are to a greater or lesser extent sclerified, the most pronounced cases are evident in some of the West Indian taxa of section *Polymorphae* (Pl. V, figs. F and G).

It is very likely that a more comprehensive and detailed study of venation in *Jatropha* will provide good taxonomic criteria that are not apparent at the present time. An example of tax-onomic distinction based on venation (as well as on leaf shape) may be seen in the disputed status of *J. integerrima* and *J. hastata* (Pl. V, figs. E and F); although McVaugh combined the two species on the basis of excessive foliar variability, the latter has recently been considered sufficiently distinctive to be made a variety of *J. integerrima* by Fosberg (1976).

Palmate venation patterns occur in broad leaves irrespective of lobing, while pinnate vena-tion is characteristic of several species with ovate-lanceolate to narrowly lanceolate leaves (Pl. V, fig. G; Pl. VI, fig. A; Pl. VIII, fig. E).

D. Petiole

The taxonomic significance of the petiole lies not only in its length—from nearly sessile (section *Mozinna*) to longer than the blade (section *Curcas* and section *Peltatae*)—but also in the presence (e.g. *Jatropha* subsection *Adenophorae*) or absence of glands. A characteristic groove runs from the base to the point of attachment of the blade in many species (e.g. *J. standleyi*). Anatomical features, particularly the number of vascular bundles, are quite con-stant for sections and/or subsections (Miller and Webster, 1962; Dehgan, 1976; Dehgan, in press).

E. Stipules

Stipules, because of their diversity, may be considered taxonomically important in sub-sectional delimitations, and less so in sectional divisions. While they are well developed in species of some sections, in others they are vestigial or highly reduced. For example, in sec-tion *Peltatae* stipules are stiff-glandular and/or branched (*J. podagrica*–Pl. X, fig. C) or they may be long and filiform-divided as in *J. multifida*. Stipules of section *Jatropha,* on the other hand, are characteristically glandular and may or may not be branched (*J. gossypiifolia*–Pl. X, figs. J and K), or may be filiform and gland-tipped (*J. velutina*). Reduced and/or early de-ciduous stipules occur in section Polymorphae (*J. integerrima*–Pl. X, fig. E). In subgenus *Cur-cas,* although most species are stipulate, variations are considerable; the glandular stipules may range from undivided to multifid (*J. cordata*–Pl. X, fig. I), single filiform, but considerably reduced (*J. dioica*–Pl. X, fig. H), or may be early deciduous (*J. curcas*–Pl. X, fig. D). Spinose stipules are completely absent from the American taxa, but do occur in African species from the Somali Republic and Arabia (Pl. X, fig. G), all of which have been placed by Pax (1910) in section *Spinosae.*

4. GLANDS

The nectar-secreting floral and extrafloral glands of *Jatropha* are significant in the taxono-mic treatment of the genus. For example, section *Jatropha* has been described as having stipi-tate glands on the leaf margins, calyx lobes, stipules, petioles, and bracts (Pax, 1910; McVaugh, 1945b). A majority of the species in all other sections are also more or less prominently gland-ular, with a number of species being glandular only at the seedling stage (*J. integerrima*) or on young growth following dormancy (*J. cinerea*). The following brief discussion is intended to show the position and distribution of the extrafloral glands in the genus.

Four distinct extrafloral gland systems can be recognized:

1. Laminar. Of all the species studied, very few show no foliar glands at any stage in their life cycle (viz. *J. hernandiifolia, J. unicostata, J . capensis,* and *J. macrorhiza*). The laminar stipitate glands are otherwise common to a greater or lesser extent in various species. These glands are most prominent in sections *Jatropha* and *Loureira.* Laminar glands may be absent in mature leaves of some taxa within a section, but present in others; e.g. *J. podagrica* and *J. multifida* lack the glands that are so prominent in *J. augustii* of section *Peltatae.* Foliar glands may be present only at the seedling stage but completely absent at maturity, as in taxa of section *Mozinna (J. dioica).* Foliar glands in some species may be seemingly sessile, as in *J. vernicosa* and *J. paradoxa,* but in fact have very short stalks.

2. Petiolar. The presence of glands on the petiole is characteristic of a number of species in section *Jatropha* subsection *Adenophorae,* but also to some extent of *J. standleyi* in section *Loureira.* These are usually long-stalked glands that may be branched, unbranched, or both as in *J. gossypiifolia* (Pl. X, figs. J and K).

3. Stipular. Glandular stipules are characteristic of a number of species in section *Jatropha* and several taxa in section *Peltatae,* but also occur on some species of section *Loureira* (e.g. *J. cordata*–Pl. X, fig. I). Filamentous and frequently deciduous stipules that are present in many species in the genus may have secondarily lost their glandular tips; stipules become long and branched in several species which are related to those with glandular stipules.

4. Bracteal. Glands of the bracts of species in section *Jatropha* are stipitate, but glands of bracts in subsection *Loureira* are covered with stipitate or sessile glands on the margins.

Generally, no evolutionary pattern can be shown for the presence or absence of extrafloral glands. Their occurrence, therefore, might be (in *Jatropha macrorhiza*) functionally related to pollinators (e.g. ant pollination) as suggested by Ehrendorfer (1973) in regard to extrafloral glands of Malvaceae and Malpighiaceae. The term extrafloral nectaries is consequently appropriate, particularly since secretion of sugary substances is a common feature of these glands (see below).

Excluding the stipular glands, all other extrafloral nectaries are marginal and none occurs (except in *J. paradoxa*) on the leaf surface. The glands of the leaf blade, bracts, and sepals are never branched, whereas those of the petiole and stipules may be branched or simple, as indicated above. Regardless of the degree of branching, the glands are generally knob-shaped, multicellular structures with more or less flattened tips and with a single layer of tightly arranged parenchymatous cells (Pl. X, fig. L). Extrafloral glands of *J. gossypiifolia* have been studied by Untawale and Mukherjee (1969). They indicated that the laminar glands originate from initials which are derived from several protodermal and subprotodermal cells, while petiolar and stipular glands originate from the development of bicelled, pointed trichomes. The laminar glands receive their vascular supply from the marginal strands of the leaves. The supply is usually unistranded and terminates just below the tip of the glands (Kakkar and Paliwal, 1972; Untawale and Mukherjee, 1969). A similar condition is reported by Bernhard (1966) for *J. zeyheri* and *J. gallabatensis* (Pl. VI, fig. B). Generally, the above description seems to be true of all the species studied here, and variations are apparently confined to the length of the stalk and the width of the tip.

The vascular supply of petiolar glands is provided by the two supernumerary bundles that traverse the petiole and are evident in the cross-sections of *J. gossypiifolia* petiole.

5. INFLORESCENCES

Any consideration of the morphology and evolutionary trends of the inflorescence in *Jatropha* must include a brief review of terminology. Much discussion in recent years has been devoted to the typology of inflorescences (for reviews see: Rickett, 1944, 1945; Weberling, 1965; Foster and Gifford, 1974; Stebbins, 1974). According to Troll's terminology (1964), the inflorescence in *Jatropha* is the "monotelic" type; the apex of the inflorescence axis ends with a terminal flower. Repetitions of the structure of the main axis are referred to as "paracladia." "Co-florescences" also occur in addition to "main-florescences" of the main axis. Consideration should also be given to the principle of variable proportions as proposed by Weberling (1965), whereby the different elements of a monotelic inflorescence may vary in many quantitative respects. The system of paracladia may be modified by the shortening and lengthening of the different internodes or by increasing or decreasing the ramification—the same as amplification and reduction of Stebbins (1974). The paracladia may be so far reduced that the terminal flower alone remains at the apical or lateral position on the flowering shoot. The main features of the type are apparently always retained (Troll, 1964; Weberling, 1965). Evolutionary trends in *Jatropha* can now be explained according to these principles.

The evolutionary trends in the genus have apparently followed two distinct pathways. On the one hand, modification and ramification of coflorescences and paracladia have resulted in the formation of highly symmetrical, compound dichasia in subgenus *Jatropha* and, on the other hand, inflorescences have become so drastically reduced as to result in a few or solitary terminal or lateral flowers in subgenus *Curcas*. In both instances the position of the inflorescence has shifted from a lateral to a terminal position.

Section *Curcas* (and *J. curcas* in particular) presents the least modified inflorescence in the genus (Pl. XII, figs. A-C), where, in addition to the main florescence, there exists a distinct coflorescence. The paracladia of each branch form a tight inflorescence with segments (dichasia) ending in a pistillate flower, while lateral branches of each dichasium produce several staminate flowers. The inflorescence of section *Peltatae* (Pl. XI, figs. A-D) presents perhaps the most perfect, typical compound dichasium. A staminate flower nearly twice as large as other flowers of the inflorescence appears at the apex of the main florescence. Di- or trichotomous paracladia form several dichasia, each of which is subtended by a bract and ends in a pistillate flower.

Section *Polymorphae* (Pl. XIII, figs. A and B) also has only a main florescence but, because of the close proximity of other inflorescences on the flowering shoot immediately below it, these appear to occupy the position of a coflorescence. Close examination reveals that all of the inflorescences are lateral although they appear terminal for a short period of time before the shoot meristem resumes growth, and should more properly be referred to as subterminal. Although the inflorescence as a whole seems tightly arranged, the paracladia are rather far apart from each other as compared to section *Peltatae;* paracladia continue growth until the inflorescence appears as a typical compound dichasium which, because of the larger terminal pistillate flowers, forms an umbrella-shaped inflorescence (e.g. *J. integerrima* Jacq.—Pl. XIII, figs. A and B).

Section *Collenucia* (Pl. XIV, figs. C and D) represents the end of the evolutionary line in the subgenus *Jatropha.* The inflorescence is generally reduced in size, length, and number of flowers, but certain features of section *Curcas* have been retained (e.g. presence of coflorescence which has been reduced to a single staminate flower in at least two of the African species, *J. marginata* and *J. paradoxa*). The position of the inflorescence is generally terminal although, as indicated earlier, there is no consistent relationship between inflorescence and formation of lateral shoots. Acrotonic ramification has resulted in reduction of the number of dichasia and, consequently, the number of flowers per paracladium.

Transitional conditions are usually found among sections, so that a continuous series can be shown. Some exceptions to the general trend of reduction occur in the subgenus *Jatropha,* however, usually as a result of extreme reduction in dichasia (e.g. *J. hernandiifolia* in section *Polymorphae*–Pl. XIII, figs. E and F) or length of the rachis (e.g. *J. capensis* in section *Tuberosae*–Pl. XI, fig. F). Other exceptions, such as terminal inflorescences in a section with an otherwise characteristically subterminal position (as in section *Polymorphae*), represent an abrupt shift and are more difficult to explain. In the case of *J. macrorhiza* (Pl. XIII, fig. C) and other geophytic species, however, consideration should be given to ecological adaptation and growth habit. Under extreme xeric conditions it is advantageous for the plants to grow and produce the largest number of flowers in the shortest possible time before the moisture supply is exhausted and the heat becomes prohibitive. Once the terminal inflorescence on the main branch is initiated, several lateral shoots grow which soon terminate in smaller inflorescences.

Difficulties arise in discussing subgenus *Curcas* because the inflorescences appear to be lateral in many species (Pl. XV, figs. D and E); in others they may be terminal and racemose-paniculate (compound dichasium–Pl. XV, figs. B and C), or the flowers are few or solitary in axils of foliage leaves or appear to be in a terminal position on the axis. The problem is compounded by the male and female inflorescences and flowers in dioecious species which have often evolved in different directions, as pointed out by Webster (1956, p. 229). However, except in cases where dioecism is absolute and plants invariably produce only pistillate or staminate flowers, as in section *Mozinna* (Pl. XVI, figs. A-C), inflorescences can be changed by manipulation of environmental conditions. For example, short days result in production of staminate flowers, while long days cause either a drastic increase in or a complete change to pistillate flowers (Dehgan, personal observations). Hybrids between otherwise monoecious species may become dioecious due to abortion of the flowers of one sex or the other (e.g. *J. curcas* × *multifida*), or else a complete change in sex expression occurs with no apparent abortive flowers (e.g. in an individual plant of *J. cathartica* × *podagrica*). The explanation for such changes probably lies in hormonal control of sex expression. Internal or external conditions may cause an imbalance of auxins which result in a shift from monoecism to dioecism, or possibly vice versa (e.g. as shown in *Cucumis* by Galun et al., 1964). However, it has not yet been possible to induce any of the normally dioecious species of *Jatropha* to become monoecious, although treatment with auxins could result in such a change, as has been reported by Heslop-Harrison (1959, 1964) for *Cannabis.* Once a species of *Jatropha* becomes dioecious, the condition is apparently irreversible. Inflorescences of such taxa proceed to evolve independently (usually by reduction in number and increase in size of pistillate flowers) and in different directions. The above explanation is pertinent not only in consideration of evolutionary trends of inflorescences and flowers, but also in the taxonomic treatment of closely related monoecious and dioecious species and their inclusion in the same section.

Inflorescence evolution within subgenus *Curcas* begins with the elaborate aggregation of flowers in *J. curcas* (section *Curcas*) as discussed above. Section *Platyphyllae* includes both monoecious and dioecious species showing reduction in number of flowers and length of the pedicel (Pl. XV, fig. A). In *J. platyphylla* the staminate inflorescence is large and has numerous flowers, but the general dichasial character has been retained, in spite of the absence of pistillate flowers. This is contrary to the "monotelic" definition of Troll (1964) since the inflorescence does not end with a terminal flower; but because the branching of the inflorescence is homologous to those inflorescences terminating in a pistillate flower, the term "polytelic" is not appropriate (cf. Weberling, 1965). The pistillate inflorescences are reduced to a few flowers, each of which terminates in a paracladium consisting of a short branch about 2 cm long. Both staminate and pistillate inflorescences appear terminal on the shoot apex. The apical meristem does not seem to have been displaced, hence, continuation of growth is by means of the lateral bud in the leaf axil (Pl. XV, fig. A).

Dioecious and monoecious species also occur in section *Loureira.* Further reduction, however, has taken place in the inflorescence of both sexes, particularly in the number of pistillate flowers. In *J. cordata* a solitary, large, pistillate flower (which usually abscises within 24-48 hours) terminates the inflorescences, with the staminate flowers occurring in loose cymes that far surpass the pistillate flower (Pl. XV, fig. C). Although no other pistillate flowers occur in such inflorescences, exclusively female plants exist that have only 2 or 3 pistillate flowers with short pedicels arising directly from the apex. The term gynodioecious can therefore properly be applied to such inflorescences. The inflorescence apparently removes the effect of apical dominance, since usually 2(3) branches appear immediately below it. A nearly identical situation exists in the *J. cinerea* group. The only difference is the absence of the single, pistillate flower in most of the taxa, in which case the plants are completely dioecious. The staminate flowers are represented by 2 or 3 individual flowers terminating the shoot apex. In a few species, including *J. canescens,* the inflorescence displaces the apical meristem, resulting in a zigzag growth form. *Jatropha fremontioides* is monoecious, but inflorescences occur as short, subsessile groups of flowers in the leaf axils with no clear distinction between the exact location of pistillate or staminate flowers. *Jatropha neopauciflora,* a related species, is dioecious and has inflorescences that are similar in position to those of *J. fremontioides.*

Plants of the section *Mozinna* show variation with respect to lateral or terminal position of the inflorescence. *Jatropha dioica* (Pl. XVI, figs. B and C) has many-flowered male inflorescences which are nearly sessile and in terminal position on short shoots, but the pistillate flowers occur singly or in groups of 2 or 3 in either a lateral position in the leaf axil or terminal (Pl. XVI, fig. A). In *J. cardiophylla* and *J. cuneata,* however, both male and female inflorescences occur on a lateral position, either as solitary flowers or in groups of 2-3(4). Variations of this have been mentioned in connection with the growth habit.

From the above discussion the following points should be re-emphasized:

1. The solitary axillary or terminal flowers are the result of reduction and, therefore, do not represent a primitive state. This is in agreement with the view of Rickett (1944), Eames (1961), and Bailey and Nast (1945). Troll (1964, 1969) has clearly demonstrated by reference to transitional forms that there is no fundamental difference between terminal racemes and a series of solitary flowers in the leaf axils. Their varied appearance is due to a more or less progressive reduction of the bracts within the inflorescence.

2. Although the inflorescences of *Jatropha* are much less complex than many other genera in the Euphorbiaceae, they nonetheless present a broad spectrum of diversity, and can be shown to be related to one another through intermediates. This for the most part is apparent in Plate XI through Plate XVI, and in Figure 2 where various stages of reduction and modification are indicated. A more complete picture cannot be presented here because less than one-third of the species have been studied as living plants.

3. No amplification through the increase of the number of branches has occurred at any stage in the evolution of the inflorescences in the genus. This can easily be determined when the number of paracladia in *J. curcas,* the most primitive taxon, is compared to those of other species. Excluding length of pedicel in section *Peltatae,* the reduction has invariably occurred by means of supression of branches (cf. Stebbins, 1974).

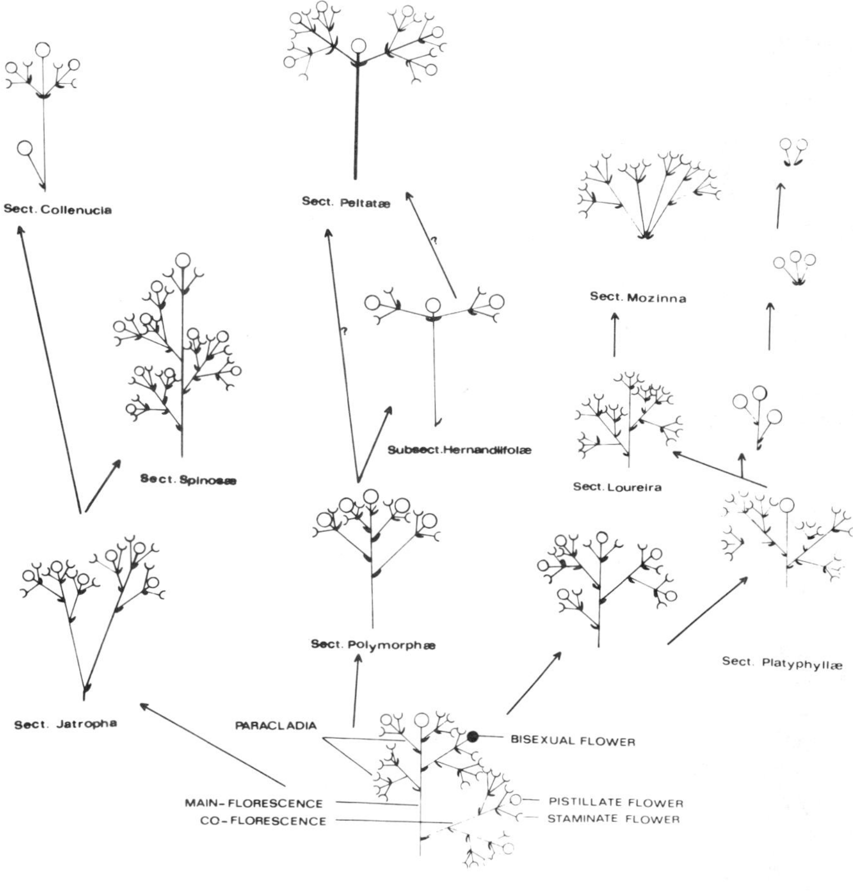

Figure 2. Evolution of inflorescence in the genus *Jatropha* L.

4. Reduction of inflorescences to a few flowers or to a single flower is probably due to the water economy brought about by overall reduction in size of the plant, leaves and reproductive parts in *Jatropha,* as well as in other xerophytic genera.

6. FLOWERS

The flowers of *Jatropha* are characteristically unisexual except for occasional hermaphroditic flowers in *J. curcas* (Venkata Rao and Ramalakshmi, 1968; Venkata Rao, 1971; and personal observations). We have not observed vestigial organs of the other sex in normal unisexual flowers of *Jatropha. Jatropha tanjorensis* has also been reported by Ellis and Saroja (1961) to have bisexual flowers (i.e., with functional staminodes in the pistillate flower). Anthers are occasionally produced in aberrant bisexual flowers of *J. gossypiifolia*, but no viable pollen is present (personal observation; Pl. XXII, fig. A). The implication of Venkata Rao (1971) that hermaphroditic flowers with viable pollen are (regularly) produced at the lower dichotomies of the inflorescence in *J. curcas,* and that flowers at higher dichotomies are male by abortion (because they show pistillodes) cannot be confirmed here. Seed-grown living plants from eighteen different sources, including India, showed only occasional bisexual flowers and never with such regularity as has been indicated by Venkata Rao (1971). One would have to agree, however, that flowers of *J. curcas* are among the least modified in the Crotonoideae and that the arborescent habit, the large, palmately lobed leaves in association with the unmodified inflorescence, and at times bisexual flowers, are sufficient grounds to consider *J. curcas* the most primitive member of the genus and fairly close to the ancestral type of the subfamily Crotonoideae. In this connection, an F_2 progeny of *J. curcas* X *integerrima* has flowers which are invariably bisexual and may indicate an expression of the ancestral feature.

The evolution of the flower, as of the inflorescence, has proceeded along two distinct lines. In subgenus *Jatropha,* there has been reduction and rearrangement of stamens with no apparent change in number of styles or carpels, while in subgenus *Curcas* the number and arrangement of stamens have remained constant, but the style branches and carpels have been gradually reduced to one. The subgeneric lines are, therefore, well demarcated on this basis.

Elucidation of taxonomic categories in *Jatropha* has historically been based on floral morphology, which seems to provide the soundest criterion for this purpose (see especially Pohl, 1827-1831; Baillon, 1858; Grisebach, 1859; Mueller, 1866; Pax, 1910; McVaugh, 1945b). Because all of the floral parts have undergone evolutionary modifications and/or reduction, they will be discussed separately.

A. Calyx

McVaugh (1945b) considered the diversity in calyx morphology significant in the delimitation of his four sections. Generally, evolution in subgenus *Jatropha* has proceeded toward a decrease in the degree of overlapping of the sepals (i.e., becoming less distinctly imbricate and with a distinct sinus between adjacent sepals). In subgenus *Curcas,* specialization has involved increasing degrees of connation of the sepals, rather than imbrication.

Serious consideration should be given to the differences in calyx lobes of staminate and pistillate flowers. The sepals of the flowers in both sexes are beset with stipitate marginal glands in section *Jatropha,* but in subsection *Loureira* glands of the sepals are more or less sessile (Pl. XXV, figs. A-D). Foliaceous sepals occur only in pistillate flowers of some species in section *Curcas* and subsection *Canescentes,* but are quite prominent in flowers of both sexes (but particularly in the pistillate flowers) in section *Platyphyllae* (Pl. XVII, figs. G and H). McVaugh (1945b) has referred to the calyx lobes of sections *Polymorphae* and *Peltatae* (*Macranthae* of McVaugh) as scarious or subherbaceous in the staminate, but scarious in the pistillate calyces.

Careful examination of calyx lobes, bracts, and stipules of *J. integerrima* reveal the presence of prominent filiform extensions with mucronate tips, a condition that is quite pronounced in *J. macrorhiza,* where sepals, stipules, and bracts are filiformly branched. Inclusion of this species in section *Polymorphae* can, therefore, be justified on this and other floral characters as well as on the crossability of the two species (Dehgan, 1976). Removal of *J. hernandiifolia* and *J. divaricata* from section *Mozinna* of McVaugh and their inclusion in section *Polymorphae* is also justifiable on the basis of the similarity of calyx lobes, among other characters.

B. Corolla

In past treatments of the genus, degree of coherence of the petals has been used to recognize genera (Adanson, 1763; Pohl, 1827-1831; Baillon, 1858) or subgenera and sections

(Grisebach, 1859; Mueller, 1866; Pax, 1910). As pointed out by McVaugh (1945b), a single corolla character is too simple to form the basis of a natural classification. To show phylogenetic relationships, the totality of characters as well as eco-geographical similarities and differences would have to be considered. The fact remains, however, that together with other floral characters, coherence of the petals provides a remarkable clue in establishing lines of divergence in the genus.

In general, evolutionary trends are along two distinct lines; in subgenus *Jatropha* the primitive condition of coherence at the base of pubescent petals (in *J. curcas*) has been maintained in section *Jatropha*. In section *Peltatae* (Pl. XIX, figs. A-L) the petals are quite distinct with a broad sinus between them. Species in section *Polymorphae* (Pl. XVIII, figs. A-J), with imbricate aestivation of the petals, occupy an intermediate position between the other two sections.

Once again, inclusion of *J. macrorhiza* in section *Polymorphae* can be justified, because, contrary to McVaugh's statement (1945b, p. 281), the corolla is not tubular and the arrangement of the petals is identical to that of *J. integerrima* and its related species (cf. Pl. XVIII, figs. A-D). One would have to sympathize with McVaugh, however, since *J. macrorhiza* is among those species with nyctonastic and/or thermonastic habit, where flowers are open for only a few hours during the warmest part of the day (10:00 A.M. to 4:00 P.M.). Since botanists understandably avoid collecting desert plants during these hours because of the extreme heat, most specimens are probably collected early in the morning or late in the afternoon, at which time the flowers are partially closed and seemingly tubular.

Increasing connation of the petals in the African section *Collenucia* (Pl. XXIV, figs. A-D) has resulted in tubular flowers very much like those of section *Mozinna* of Mexico. Although the two groups differ in several technical characters, this corollar resemblance resulted in their inclusion in the same subsection (*Brachyblastae*) by Pax (1910) and recognition as a section in the same subgenus (*Curcas*) by Chiovenda (1930). In actuality, as shown in this paper, they share little other than common generic ancestry. Their similarity should therefore be considered as due to parallel development in two widely separated geographical localities.

A continuous series of increasing coherence can be shown to exist in subgenus *Curcas:* in section *Curcas* (Pl. XVII, figs. A-F) where petals are joined only at the base; in section *Platyphyllae* (Pl. XVII, figs. G-H) connation has reached to nearly one-third the length of the petals, resulting in more or less urceolate flowers; species of section *Loureira* (Pl. XXV, figs. A-H) in general have petals that are joined to more than two-thirds of their length to form subglobose flowers. Except for the recurved tips of the petals the connation is otherwise complete in species of section *Mozinna* (Pl. XXVI, figs. A-H) and, consequently, the flowers are tubular.

The color of the petals is commonly red in the American species of subgenus *Jatropha,* as in sections *Peltatae, Polymorphae* (except subsection *Hernandiifoliae*), and *Jatropha* subsection *Adenophorae.* The Old World taxa are usually green to yellow or yellow-brown. In subgenus *Curcas* the color is red to white, but various color combinations occur, particularly in section *Loureira.*

C. Androecium

The number of stamens in *J . curcas* is 10, arranged in two indistinct whorls of 5 each in a single column (monadelphous), and in close proximity to each other. This basic arrangement has been retained in other taxa of subgenus *Curcas,* except that the two tiers of 5 stamens each are distinctly separated (biseriate). The trend in subgenus *Jatropha,* however, has been toward reduction in number and/or loss of the staminal column. In section *Peltatae* there are

8 free stamens (e.g. *J. augustii, J. multifida,* and *J. cathartica,* etc.), and the flowers are consequently apostemonous. The stamens in *J. podagrica,* which are preponderantly 8, may frequently vary from 6-14 in the same inflorescence (see Pl. XIX, fig. F, with 9 stamens); this is usually accompanied by an increase in the number of petals, and may indicate recent hybrid origin because these characters are common in artificial hybrids.

The number of stamens in the West Indian species of section *Polymorphae* (Pl. XVII, figs. G and H) is 10 with a monadelphous-biseriate arrangement; but reduction to 8 has occurred in *J. macrorhiza* (subsection *Macrorhizae*) (Pl. XVII, figs. C and D), as well as in the African species of section *Tuberosae* (e.g. *J. capensis, J. gallabatensis,* and *J. unicostata*) (Pls. XX and XXI). In these species two whorls of stamens are present with 5 stamens in the lower and 3 in the upper whorls. The situation is similar in section *Jatropha* (Pl. XXII, fig. B; Pl. XXIII, figs. C, F, and G), except that the number of stamens is consistently 8, and the staminal column is considerably shorter than that of section *Polymorphae* (never extending beyond the petals).

McVaugh (1945b) included species with both 8 and 10 stamens in sections *Peltatae* ("Macranthae") and *Jatropha*. Although none of the species with 10 stamens have been available for this study as living plants, the position of *J. macrantha* appears reasonably certain. Photographs taken in Peru, as well as herbarium material, indicate that it should more properly be included in section *Polymorphae* on the basis of its 10 monadelphous-biseriate stamens, aestivation of the petals, corymbose inflorescence and 3-lobed leaves. The taxonomic position of *J. macrocarpa* and *J. papyrifera* will have to await future determination because no material has been available for examination; but on the basis of McVaugh's descriptions, they should also be included in *Polymorphae*.

Inclusion of any species with 10 stamens in section *Jatropha* is dubious and, except for teratological occurrences, no such species probably belong to this section. It is notable that a question mark has been invariably used by McVaugh (1945b) whenever he referred a species with 10 stamens to section *Jatropha*. As a general rule, the Old World species of *Jatropha* have only 8 stamens; excluding *J. ferox,* no other species have ever been reported with 10 stamens.

D. Gynoecium

In accordance with other genera in the subfamily Crotonoideae the gynoecium in the majority of species in *Jatropha,* including all of subgenus *Jatropha* and a number of taxa in subgenus *Curcas,* consists of three united carpels with a solitary ovule in each locule of the ovary. Phylogenetic lines, however, can be decisively established within the genus on the basis of a series of reductions that have occurred in the Mexican species (subgenus *Curcas*). In *J. curcas* (section *Curcas*) the three slender styles are connate to about two-thirds of their length, dilating to massive bifurcate stigmata (Pl. XVII, fig. A). The three styles are free, but have retained the bifid character in section *Platyphyllae* (Pl. XVII, fig. G). In section *Loureira* one of the styles is lost (with exceptions in some individual plants of the *J. cinerea* group) and the other two are joined to form a stylar column which dilates only at the tips to form the two bifid stigmata (Pl. XXV, figs. A, B, E, and F). Reduction of the gynoecium reaches its ultimate evolutionary stage in section *Mozinna* where only a solitary bifurcate style remains (Pl. XVIII, figs. A, B, E, and G).

Sections *Peltatae* and *Jatropha* differ only with respect to the length of the stylar column, which is nearly twice as long in section *Jatropha* (cf. Pl. XIX, figs. A and B; Pl. XXII, figs. B and C). The stylar column, however, is completely absent in section *Polymorphae* (Pl. XVIII, figs. A, B, E, F, and I), and in subsection *Capenses* of section *Tuberosae* (Pl. XX, figs. A, B, E, and F), and styles arise independently from the apex of the ovary. In this connection it should be noted that both *J. macrorhiza* and *J. macrantha* (see drawing in McVaugh, 1945b, p. 275) have free styles that differ only slightly with respect to the shape of the stigma lobes and length of the style.

The vasculature of the gynoecium and of the flower in general is apparently uniform within the genus and is of little taxonomic value. Therefore, it will not be discussed here. For details the reader is referred to Rao and Ramalakshmi (1968), Nair and Abraham (1962), Froembling (1896), Beille (1902), Gaucher (1902), Michaelis (1924), Solereder (1908), and Metcalf and Chalk (1950). Embryology of *J. curcas* and *J. gossypiifolia* has been studied in detail by several authors (Arnoldi, 1912; Kajale and Rao, 1943; Singh, 1970). They found that both species are nearly identical in having bitegmic and anatropous ovules. The slender nucellar beak projects beyond the integuments and lies in close contact with the obturator. Vascular strands are present in the inner integument and the hypostase is distinct and persists in seeds. Embryology of *J. glandulifera* and *J. hastata* has been described by Wiehr (1930—from Singh, 1970), and is apparently similar to the other species. Preliminary examination of these and a few other species by us confirms their findings; excluding minor differences, the ovule seems quite uniform throughout the genus, including the derived Mexican species.

E. Pollen

The pollen grains in genus *Jatropha* are typically Crotonoid, and uniformly spheroidal, inaperturate, and with hexagonally arranged exinous knobs (Erdtman, 1952; Punt, 1962; Lynch and Webster, 1976). The knobs differ in section *Peltatae* by being smooth and lacking the typical striations (Pl. XXVIII, figs. A-E) that are found in nearly all the other taxa (Pls. XXVII, XXIX). Minor knobs which are found in some pollen grains (see, for example, Pl. XXVIII) have thus far proven not to have any taxonomic significance in *Jatropha* because they are inconsistent both in their occurrence and number even within a single taxon. Extreme difficulty was encountered in working with whole grains due to charging. Therefore, frozen sections of the pollen grains, as described by Muller (1973), were used to show the details of the exinous knobs.

Differences in diameter of the pollen grains along with reductions in the reproductive structures have occurred from section *Curcas* ($\bar{x}$ 83.5 μ in *J. curcas*) to section *Mozinna* ($\bar{x}$ 69 μ in *J. dioica*) with intermediate sizes present in the other two sections (*Platyphyllae* and *Loureira*) of the subgenus *Curcas* (Dehgan, 1976). The range of variation in size of the grains in subgenus *Jatropha*, however, is much narrower and is closer to *J. curcas* ($\bar{x}$ 79 μ in *J. macrorhiza* to 86 μ in *J. integerrima*). If *J. curcas* is accepted as the most primitive taxon in the genus, then reduction in size in subgenus *Curcas* can be explained to be the result of divergent lines of evolution in the two subgenera. Reduction in pollen-grain diameter seems to be directly correlated with south to north distribution of the taxa and, consequently, to increasing aridity.

Only two nuclei were reported by Webster and Rupert (1973) in the pollen grains of the four species they examined. Two other species, *J. capensis* and *J. dioica*, are also binucleate (Dehgan and Craig, unpublished data). The probability that the pollen grains in the genus are all binucleate presents further evidence for the naturalness of the genus.

7. FRUITS

In South American, West Indian, and Old World species of *Jatropha* the fruit is a trilocular, explosively dehiscent capsule with a single seed per locule. In the great majority of Mexican taxa, however, the fruit has undergone reduction in conjunction with the stigma lobes as discussed above.

Jatropha curcas has a trilocular, ellipsoidal, subdrupaceous fruit (Pl. XXX, fig. H), the exocarp of which remains fleshy until the seeds are mature, finally separating into three cocci. The fruit is 2.5-3.5 cm long and 2-2.5 cm wide. The fruits of *J. platyphylla* and related species are variable in size (1.5-3.5 cm long and wide) and trilocular, but with a deep depression between the locules. Therefore, these fruits are distinctly trilobed, a condition which is apparent in the early stages of ontogeny (Pl. XVII, figs. G and H). The fruit is capsular, but not explosively dehiscent. In section *Loureira* there are two locules (Pl. XXX, fig. L), and in section *Mozinna* there is a tardily dehiscent, unilocular fruit 1.0-1.5 cm long and wide with a single seed (Pl. XXX, figs. J and K). Transitional conditions may occur; for example, in section *Loureira* there is at least one species with mainly trilocular fruit (*J. giffordiana*) and unilocular fruits (by abortion) may occur; and in section *Mozinna* bilocular fruits may occasionally be produced by the usually unilocular *J. dioica.*

Fruits in subgenus *Jatropha* are consistently trilocular (Pl. XX, figs. A-G). There is a gradual reduction in thickness of the mesocarp and in size (from 1.5 to less than 1 cm, but with a few exceptions, e.g. *J. multifida,* which is 3-4 cm long and wide). Section *Collenucia* has the smallest fruit and the thinnest wall, which dehisces violently at maturity.

The fruit is green in all species, but becomes yellowish at maturity and just before dehiscence. The fruit is often pubescent in subsection *Adenophorae* of section *Jatropha* (Pl. XXX, fig. A), and in section *Collenucia.* Fruits of the species in other sections are glabrous.

8. SEEDS AND SEEDLINGS

One of the principal distinguishing characters for establishing the phylogenetic lines in the two subgenera is the shape of the seed and the size and morphology of the caruncle.

Seeds of *J. curcas* are black, oblong, 2.5-3 cm long, and 1 cm in thickness (Pl. XXXIII, fig. A). The relatively small caruncle is appressed to the beak, and is distinctly lobed. More or less spherical or ellipsoidal seeds with a very small caruncle characterize all other species of subgenus *Curcas* (Pl. XXIII, figs. B-D). The only exception, according to McVaugh (1945b), is *J. fremontioides* (not seen), whose seeds resemble those of section *Peltatae* (which has the largest caruncle in the genus); for example, in *J. hieronymii* (Pl. XXXII, fig. A) the caruncle is 8-10 mm X 4-6 mm with distinct lobes. In general, the seeds in subgenus *Jatropha* are grayish-brown, mottled, and with a massive caruncle in proportion to the seed size (Pl. XXXII, figs. A-J). The smallest seeds are in the African species of sections *Jatropha* and *Collenucia.*

The embryo is of the spathulate type and four seminal roots are apparent on the hypocotyl-root axis; in seedlings, these appear as four laterals with the primary root at the center. The seed coat, according to Singh (1970), is formed by both integuments. The copious starchy endosperm is of the nuclear type (Gram, 1895-96; Mandl, 1926; Singh, 1970; Corner, 1976); it has recently been reported to be capable of producing triploid roots and shoots in vitro, indicating its morphogenetic potentiality (Srivastava, 1971).

Jatropha multifida presents the largest known seeds in the genus, which are exceptional in the subgenus *Jatropha* in being spherical (Pl. XXXII, fig. C), and have been a matter of curiosity since Holm (1899) wrote about their peculiar mode of germination. Germination of *J. mul-*

tifida and of *Hevea brasiliensis* (Duke, 1969) is unique in the Euphorbiaceae. The cotyledons are never completely freed from the seed coat (termed cryptocotylar by Duke, 1965). This situation poses a great deal of difficulty in germination of hybrid seeds where *J. multifida* is one of the parents because the cotyledons are never completely freed and, consequently, the food reserve becomes exhausted. Since photosynthesis does not begin until the first leaves appear, the seedlings are usually too weak to survive. This may perhaps explain the unusually large seeds of both *J. multifida* and *Hevea brasiliensis,* where sufficient food can be stored to compensate for the lack of photosynthesis by the initial leaves.

Germination in all other species is phanerocotylar (i.e., cotyledons are freed from the seed coat), as in other Euphorbiaceae (Duke, 1969), where the cotyledons are slightly longer than broad (Pl. XXXI). Eophylls in species of all sections except *Peltatae* are much longer than broad, never lobed, and are almost indistinguishable from species to species at this stage. Seedlings of section *Peltatae* have reniform cotyledons and lobed eophylls. *Jatropha podagrica,* with peltate eophylls, is also exceptional. In section *Jatropha* the first eophyll is entire, but the second is invariably lobed (e.g. *J. gossypiifolia*).

The cotyledonary veins are three in section *Polymorphae,* but in all other species are five, with three prominent and two weaker laterals arising from the base. The cotyledons are pseudopalmately veined (terminology of Bailey, 1956).

9. POLLINATION

To our knowledge there are no reports in the literature regarding pollination mechanisms in *Jatropha* or, for that matter, in most members of the Euphorbiaceae. Observations on plants grown at Davis and wild populations of *J. macrorhiza,* however, provide some insight into the mode of pollination in the genus.

Jatropha curcas appears to be a moth-pollinated species because of its sweet, heavy perfume at night, greenish-white flowers, versatile anthers and protruding sexual organs, copious nectar, and absence of visible nectar guides (Faegri and Van der Pijl, 1970). An infestation of the plants in the greenhouse with larvae of Noctuid moths resulted in nearly all pistillate flowers setting viable seed. Excluding the occasional hermaphroditic flowers which self-pollinate, seed set otherwise never occurs except from hand pollination.

Plants of *J. integerrima, J. podagrica,* and *J. multifida* all have conspicuous bright red flowers, free petals, and copious nectar and are probably pollinated by butterflies. The two species of butterflies observed on the flowers were apparently responsible for pollination and seed set when plants were grown outdoors in Davis during the summer. A note of interest, for which no explanation can be offered at this time, is the visitation of flowers of *J. integerrima* by the yellow race of cabbage butterflies, but not by the white race which was frequently in the vicinity.

The greenish-white flowers of *J. capensis* and *J. hernandiifolia,* are visited in Davis by a small wasp which lingers on the flowers for long periods. In both species seed set was frequent. This visitation by the same wasp is surprising because *J. capensis* has recurved petals while *J. hernandiifolia* has stiff and upright ones.

Jatropha gossypiifolia is visited by a small sweat bee (Halictidae) and midges, which are apparently instrumental in pollination. The multiple number of branches and, consequently, the numerous inflorescences and flowers in combination with self-compatibility and facultative annual growth habit, can explain the weedy nature of this species.

Jatropha macrorhiza is the only species observed in the field. Three insects were seen on the flowers: an *Andrena* bee, a sweat bee (Halictidae), and a number of relatively large ants. The possibility of ant pollination cannot be ruled out because:

1. Although there is a lag of 24-48 hr in anthesis of the staminate flowers after opening of the pistillate flowers, both types of flowers are open on the second or third day, and were open in nearly all inflorescences at the time of these observations.

2. Movement from flower to flower on compact inflorescences poses no difficulty for the ant, or for the other walking insects, particularly since the plant is often no more than 20-30 cm tall (Hickman, 1974).

3. The basic requirement for ant pollination according to Hocking (1975) is self-compatibility, which is the normal condition for all monoecious *Jatropha* spp. under consideration here, including *J. macrorhiza.*

4. Physical barriers, such as hairs (in *J. macrorhiza*), that would prevent the ants from coming in direct contact with the anthers and stigmata are lacking.

The four requirements for ant pollination mentioned here are even more applicable to all the American and African species of section *Jatropha.* The extrafloral nectaries in this section (completely absent in *J. macrorhiza*) should provide even more incentive for ant visitation.

In general, *Jatropha* is diclinous (terminology of Percival, 1965) with both monoecious and dioecious species. All taxa studied so far are self-compatible when hand-pollinated. Selfing otherwise occurs, but very rarely under cultivated conditions of the greenhouse or outdoors, indicating the possibility that geitonogamy and xenogamy are the rule under natural conditions. Xenogamy is accomplished in two ways: the normal flowering habit is such that pistillate flowers on a plant reach anthesis from one to several days earlier than staminate flowers. This arrangement results in biparental reproduction (Grant, 1975), with a gradual reduction in the number of flowers of one sex or the other finally leading to dioecism, and consequently to truly obligate outcrossing. The transitional stage is represented by such species as *J. cordata* where only a single pistillate flower has remained in an otherwise male inflorescence (gynodioecious).*

From an evolutionary standpoint, indications of an ancestral hermaphroditism and its consequential selfing are found in *J. curcas.* Diversification in floral mechanism within subgenus *Jatropha*, a monoecious group, has not been accompanied by self-incompatibility. Modification in floral structure from pleomorphic (Leppik, 1969) in *J. curcas* or a progenitor thereof to stereomorphic (the same as bell/funnel type of Faegri and Van der Pijl, 1970), where stamens, pistils, and particularly the nectar are hidden in a tube, has occurred. Whereas the nectaries of the flowers in subgenus *Jatropha* are usually exposed and accessible to flies, wasps, etc., those of subgenus *Curcas* are hidden and nectar is available only to insects with a long proboscis or, most likely, to long-tongued bees. The final modification has been attainment of the dioecious condition, thereby securing an alternative route to self-incompatibility.

Evolutionary literature is filled with examples of self-pollinating (autogamous) species being derived from xenogamous ancestors (see Stebbins, 1950, 1957, 1974). *Jatropha*, however, exemplifies a case where the opposite has occurred, i.e., the trend has been from partial self-pollination to total xenogamy. In fact, pollination mechanisms discussed here exemplify a situation where the best of both worlds has been achieved: maintenance of heterozygosity

---

*The term gynodioecious is here used in a sense somewhat different from that of Percival (1965); the population of *J. cordata* consists of pistillate plants and bisexual plants (not plants with bisexual flowers, as given in most definitions).

by means of preferential or obligate outcrossing, but assuring, as well, perpetuation of the species through self-compatibility.

Allard (1965), with regard to colonizing species, has pointed out that "there is, however, one feature which the great majority of these notably successful colonizers share: a mating system involving predominantly self-fertilization," a view which is held by most evolutionists (for reviews see Stebbins, 1958, 1970, 1974). Here again *Jatropha* offers an exception where colonization of disturbed habitats often takes place by dioecious species. As a case in point, *J. cinerea* (*sensu lato*) and *J. cuneata* are the most commonly encountered species in Baja California and on the west coast of Mexico. Within one year after completion of the main highway in Baja, these two species along with pioneering annuals have been the first plants to colonize both sides of the road, a behavior not typical of perennial woody plants. The same may be said of *J. dioica* in the Chihuahuan desert, *J. cardiophylla* in Arizona, and other species of *Jatropha* which are frequently encountered in Mexico. The explanation perhaps lies in water runoff from the road, since these areas are marginal to the main population.

10. CHROMOSOME NUMBERS

Hans (1973), in a chromosomal conspectus of the Euphorbiaceae, listed a total of nine species of *Jatropha,* of which one (*J. hastata*) is considered by McVaugh (1945b) to be synonymous with *J. integerrima.* As indicated in an earlier paper (Rupert et al., 1970), identification of individual chromosomes in *Jatropha* is hindered by their small size—from less than 2 to 4 μ—as well as by difficulty in staining. This can perhaps explain the lack of reports for other species in the genus. The notable absence of any counts for the North American species of *Jatropha* in a recent paper of Urbatsch et al. (1975) is a likely testimony to this difficulty.

Counts for somatic chromosome numbers of thirty-four species of *Jatropha* (Table 1) were made from alcoholic-aceto-carmine-HCl squashes (Snow, 1963) of young leaves. Ferric ammonium sulfate (Capinin and Bruce, 1955) was used as a mordant after staining.

The base number for the genus is x = 11. *Jatropha dioica* and *J. cuneata* are tetraploid with 2n = 44, with the latter species having more than 44 chromosomes in one collection. Due to their extremely small size, it was not possible to get an accurate count. This collection was made 9 miles N.E. of La Paz, Baja California; the plants are extremely dwarfed (less than 12 inches), with small leaves and a very deep root system (Pl. IV, fig. E).

Jatropha cardiophylla is interesting because one would expect it to be a polyploid (cf. Stebbins, 1971) on the basis of its rhizomatous growth habit and its northerly geographical distribution. But instead, it has doubled the size rather than the number of chromosomes. Although level of ploidy is not significant as such, placement of the three species, namely *J. cardiophylla, J. cuneata,* and *J. dioica* in section *Mozinna* is nonetheless warranted, in addition to cytological grounds, on the basis of morphological and anatomical similarities as well as geographical distribution.

Efficient vegetative reproduction, particularly by means of rhizomes, is correlated with a high percentage of polyploidy according to Stebbins (1971). It is, therefore, not surprising to find that polyploid (or genetically equivalent, as in *J. cardiophylla*) species are rhizomatous and ecologically aggressive. Accordingly, such highly successful rhizomatous polyploids may have eliminated their diploid ancestors through competition (Stebbins, 1971), especially if their ancestral species were not well adapted to the cold and aridity that these species now endure in their northerly range. In light of the morphological reduction series that was discussed in connection with reproductive structures in subgenus *Curcas,* it would be unwise to

Table 1
SOMATIC CHROMOSOME NUMBERS

TAXON	n	2n	AUTHORITY	U. C. D. ACC. NO.	SOURCE
J. augustii Pax & Hoffm.		22		B74.056	Peru (U.C.B. 64.224)
J. brockmanii Hutchinson		22		B75.070	Somali Rep. (Lov. #11760)
J. canescens Benth. ex Muell. Arg.		22		B71.069	Mexico
J. capensis (L. f.) Sond.		22		B67.045	S. Africa (U.C.B. 66.712)
J. cardiophylla Muell. Arg.		22		B74.071	Arizona, U.S.A.
		22		B74.072	"　　　　"
		22		B74.203	Sonora, Mexico
		22		B75.054	"　　　　"
J. cathartica Terán & Berland.		22		B67.471	Texas, U.S.A.
		22		B75.057	"　　　"
J. ciliata Sessé & Cerv.		22		B75.051	Oaxaca, Mexico (Abbey Garden, Santa Barbara, Calif.)
J. cinerea (Ortega) Muell. Arg.		22		B74.008	Navajoa, Mexico
		22		B74.020	Baja California, Mexico
(species complex)		22		B74.022	"　　　"
		22		B74.023	"　　　"
		22		B74.027	"　　　"
		22		B74.028	"　　　"
		22		B74.033	"　　　"
		22		B74.035	"　　　"
		22		B74.038	(Magdalena Island)
		22		B74.195	Sonora, Mexico
		22		B74.197	"　　　"
		22		B74.199	"　　　"
		22		B74.201	"　　　"
J. cordata (Ortega) Muell. Arg.		22		B72.129	Sonora, Mexico
		22		B74.006	San Luis Potosí, Mexico

Table 1 (continued)

TAXON	n	2n	AUTHORITY	U. C. D. ACC. NO.	SOURCE
J. cuneata		44		B74.026	Baja California
Wiggins & Rollins		44		B74.027	" "
		+44(?)		B74.037	" " (highly dwarfed)
		44		B74.200	Kino Bay, Mexico
		44		B74.010	Texas, U.S.A. (probably cultivated)
J. curcas L.		22	Perry, 1943	- - -	Bengal, India
		22	Miller and Webster, 1962	- - -	- - -
	11		Datta, 1967	- - -	Bengal, India
	11		Hans, 1973	- - -	W. Himalayas
		44	Mehra and Hans, 1969	- - -	Bengal, India
		22		B68.286	Dakar, Senegal
		22		B73.128	Bogor, Java
		22		B74.011	Mayaguez, Puerto Rico
		22		B74.013	Dehra Dun, India
		22		B74.015	Lucknow, India
		22		B74.039	Georgetown, Guyana
		22		B74.014	Freetown, Sierra Leone
		22		B74.156	Papua, New Guinea
J. dioica Sessé	22		Miller and Webster, 1966	- - -	Texas, U.S.A.
		44		B71.070	Chihuahua, Mexico
		44		B74.058	" "
		44		B74.226	" "
J. excisa Griseb. var. *pubescens* Lourteig & O'Donnell		22		B74.166	Origin unknown (Cultivated, Abbey Garden, Calif.)
J. fremontioides Standley		22		B74.163	Oaxaca, Mexico (Cultivated, Huntington Bot. Gard., Calif.)
J. gallabatensis Schweinf.		22		B74.161	Somali Rep. (N.B.G. #22824)
J. giffordiana Dehgan & Webster		22		B74.019	Baja California

Table 1 (continued)

TAXON	n	2n	AUTHORITY	U. C. D. ACC. NO.	SOURCE
J. glauca Vahl		22		B75.048	Aden (ISI. 298)
J. gossypiifolia L.		22	Perry, 1943	- - -	Florida, U.S.A.
		22	Miller and Webster, 1962	- - -	West Indies
		22	Nanda, 1962	- - -	India
	11		Datta, 1967	- - -	Bengal, India
		22		B69.006	Ghana
		22		B68.144	G.L.W. 67.194 (?)
		22		B74.016	Lucknow, India
		22		B74.040	Buayana, S. Afr.
J. hernandiifolia Vent.		22	Miller and Webster, 1962	- - -	West Indies
		22		B67.281	West Indies, G.L.W.
J. hieronymii Kuntze		22		B74.068	Argentina (collected by Otto Solbrig)
J. integerrima Jacq. (as *J. hastata* Jacq.)		22	Rupert et al., 1970	B67.278	Cuba (from cultivation)
		22	Miller and Webster, 1962	- - -	Cultivated, U.S.D.A., Miami
J. macrorhiza Benth.		22		B74.075	Arizona, U.S.A.
		22		B74.040	Chihuahua, Mexico
J. malacophylla Standley		22		B71.072	Mexico
		22		B73.046	Sonora, Mexico
		22		B74.198	Guamuchil, Mexico
J. marginata Chiov.		22		B74.162	Somali Rep. (H.B.G. #29080)
J. mcvaughii Dehgan & Webster		22		B74.232	Jalisco, Mexico
J. moranii Dehgan & Webster		22		B75.052	Baja California (J.B. #680)
J. multifida L.		22	Perry, 1943	- - -	Honolulu, Hawaii (from cultivation)
		22	Miller and Webster, 1962	- - -	Cultivated, Fairchild Gardens, Florida

Table 1 (continued)

TAXON	n	2n	AUTHORITY	U. C. D. ACC. NO.	SOURCE
	11		Datta, 1967	- - -	- - -
		22		B74.157	Papua, New Guinea
		22		B75.029	Lisboa, Angola
J. paradoxa (Chiov.) Chiov.		22		B75.023	Somali Rep. (U.C.B.G. 69.113)
J. platyphylla Muell. Arg.		22		B69.300	Jalisco, Mexico
		22		B74.207	" "
J. podagrica Hook.	11		Miller and Webster, 1966	- - -	Jamaica
	11		Datta, 1967	- - -	Bengal, India
		22		B59.047	Cultivated, Origin Unknown
		22		B73.129	Bogor, Java
		22		B74.011	Mayaguez, Puerto Rico
		22		B74.232	Freetown, Sierra Leone
J. standleyi Steyerm.		22		B75.047	Mexico (Cultivated, Abbey Garden, California)
J. unicostata Balf. f.		22		B67.471	Socotra (U.C.B.G. 67.954)
J. velutina Pax & Hoffm.		22		B75.048	Kenya (Cultivated, Abbey Garden, California)

hypothesize on whether the polyploid taxa are auto- or amphiploids. In view of the absence of reports of naturally occurring hybrids in subgenus *Curcas*, it is likely that they are auto-polyploids. But since cases are known where hybrids may be indistinguishable from one or the other parent (e.g. artificial triploid hybrids of *J. curcas* × *podagrica* and *J. curcas* × *cathartica*), such an assumption may not be valid. Furthermore, as pointed out by Grant (1971, p. 241), "close morphological resemblance between a given polyploid and a given diploid is not a valid criterion of autoploidy." Whatever the mode of ploidy may be, it seems clear that an increase in the number of chromosomes is a result of selective pressure in response to the harsh environment of northern Mexico and Arizona. A highly successful rhizomatous polyploid, according to Stebbins (1971), would have a particularly good chance of eliminating its diploid ancestor by direct competition or interaction, and thus making it extinct, or by extending beyond the range of its ancestor. The reproductive advantage of the polyploids can perhaps explain the restriction to the Oaxacan area of Mexico of such species as *J. sympetala, J. neopauciflora* and others that are morphologically similar to the tetraploid species.

The polyploid (2n = 44) *J. curcas* that has been reported by Mehra and Hans (1969) is somewhat dubious as indicated in Table 1. A possible explanation is the fact that the generally small chromosomes of the genus are difficult to count, particularly if more than the expected number are present. Artificially induced tetraploid *J. curcas* (seedlings treated with 0.2% colchicine) and the triploid hybrids mentioned above are morphologically indistinguishable from their diploid parents. Consequently, if any such hybrids were to occur where two to three of these species are cultivated, then triploid hybrids are very likely. Spontaneous doubling of chromosomes in an individual plant, of course, cannot be ruled out either.

The single most important question with regard to chromosome number is, why has such a geographically widely distributed genus (Fig. 3) undergone so much morphological variability without change in basic chromosome number? Perhaps the simplest and most logical answer is that long-lived woody perennials of tropical origin must rely on a rich gene pool which chromosomal variability would hamper. Natural selection, therefore, would act on morphological rather than chromosomal variation (chromosome numbers are stable but rearrangements may have occurred). According to Kyhos (personal communication) constancy is achieved in such cases by longevity coupled with high recombination potential. This point will be discussed in detail in a later paper (see also Dehgan, 1976).

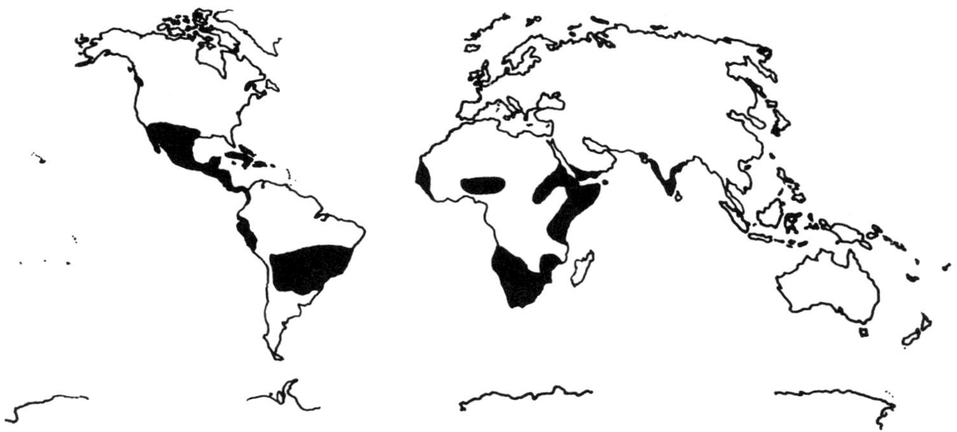

Figure 3. Geographical distribution of the genus *Jatropha* L.

A SYNOPSIS OF THE INFRAGENERIC TAXA OF *JATROPHA*

Jatropha L. [Syst. Nat. ed. 1, 1735; Gen. Pl. ed. 1, 288. 1737] Sp. Pl. ed. 1, 1006. 1753; Gen. Pl. ed.
5, 437. 1754; Endl. Gen. Pl. 2: 1114. 1840; Baillon, Étude Gen. Euphorb. 294. 1858; Griseb. Fl.
Brit. W. Ind. 36. 1859; Muell. Arg. in DC. Prodr. 15(2): 1076. 1866; in Mart. Fl. Brasil. 11(2): 485.
1874; Baillon, Hist. Pl. 5: 112, 179. 1874; Benth. & Hook. f. Gen. Pl. 3: 290. 1880; Pax in Engler &
Prantl, Natürl. Pflanzenfam. ed. 1, 3(5): 74. 1890; in Engler, Pflanzenwelt Ost-Afrikas C: 240. 1895;
in Engler, Das Pflanzenreich IV. 147 (Heft 42):21. 1910; Hutchinson in Thistleton-Dyer, Fl. Trop. Afr.
6(1): 775. 1913; Fawc. & Rendle, Fl. Jam. 4(2): 310. 1920; Standley, Contr. U.S. Nat. Herb. 23: 634.
1923; Prain in Thistleton-Dyer, Fl. Cap. 5(2): 418. 1925; Hutchinson & Dalz., Fl. W. Trop. Afr. ed. 1,
1: 297. 1928; McKenzie, Bull. Torrey Bot. Club 56: 213. 1929; Pax & Hoffm. in Engler & Prantl, Natürl. Pflanzenfam. ed. 2, 19c: 160. 1931; Lourteig & O'Donnell, Lilloa 9: 118. 1943; McVaugh, Bull.
Torrey Bot. Club 72: 31, 271. 1945; Standley & Steyermark, Fieldiana Bot. 24: 126. 1949; Alain in
León, Fl. Cuba 3: 75. 1953; Shreve & Wiggirs, Veg. Flora Sonoran Desert 1: 800. 1964; Gooding et
al., Fl. Barbados 254. 1965; Webster, J. Arnold Arb. 48: 340. 1967; Webster & Burch, Ann. Missouri
Bot. Gard. 54: 234. 1968; Correll & Johnston, Fl. Texas 953. 1970; Radcliffe-Smith, Kew Bull. 28:
283. 1973; Dyer, Gen. S. Afr. Fl. Pl. 1: 320. 1975; Allem, Rev. Brasil. Biol. 37: 209. 1977.
Syn.: *Curcas* Adans. Fam. Pl. 2: 356. 1763.
    Bromfeldia Neck. Elem. Bot. 2: 347. 1790.
    Castiglionia Ruiz & Pavon, Prodr. Fl. Peruv. 139, t.37. 1794.
    Mozinna Ortega, Nov. Pl. Hort. Reg. Eot. Matr. 8: 104, t.13. 1799.
    Loureira Cav. Icon. Descr. Pl. 5: 17, t.429. 1799.
    Mesandrinia Raf. Neogenyton 3. 1825.
    Adenoropium Pohl, Pl. Bras. Icon. Descr. 1: 12. 1827.
    Zimapania Engler & Pax in Engler, Natürl. Pflanzenfam. ed. 1, 3(5): 119. 1891.
    Collenucia Chiov. Fl. Somala 1: 177. 1929.
Lectotype species: *Jatropha gossypiifolia* L. (McVaugh, Bull. Torr. Bot. Club 71:45. 1944.)
Trees, shrubs, facultative annuals or rhizomatous subshrubs, or tuberous perennial suffrutescent herbs.
Latex universally produced, pale, cloudy, yellow to distinctly red but never milky, powdery when dry;
laticifers articulated, non-articulated, or idioblastic. Cyanogenic glucoside compounds absent (whereas they
are known in related genera*). *Roots* fleshy, with parenchymatous storage tissue, or slender; arranged in
fives, with the outer four shallow and central one deep. *Bark* of the woody species smooth, wrinkled, with
fissures and cracks, or peeling, with or without pubescence; short shoots common in the xeric taxa. *Leaves*
alternate, long petiolate to subsessile; blades simple to palmately 3-, 5-, or 7-lobed, or divided; venation
palmate or pinnate, brochidodromous (or exceptionally craspedodromous); margins entire, glandular, or
conspicuously serrate, with or without mucronate tips or glands; lamina glabrous, glaucous, or with vary-
ing degrees of pubescence or tomentose; stipules simple or branched, sometimes stiff-glandular, long filiform, filiform and gland tipped, spinose, reduced or early deciduous (rarely absent). *Inflorescence* axillary,
terminal, subterminal, or reduced to a few or solitary flowers in the leaf axils; coflorescences present and

*Unpublished data of Professor Eric Conn.

distinct, present but not distinct, or absent; paracladia few to many in compound dichasia; monoecious, gynodioecious to dioecious (flowers rarely hermaphroditic); female flowers at the proximal dichotomies and fewer than the male. *Staminate flowers* actinomorphic, urceolate, subglobose or tubular; calyx lobes 5, ± imbricate or distinct, margins entire, glandular, serrate or rarely divided, often foliaceous; corolla greenish, yellow-green, yellow-brown, white, or red, of 5 imbricate, free or coherent to variously connate petals, glabrous or with various degrees of pubescence; disc entire, dissected or lobed, of nectariferous glands; stamens usually 8 or 10 (rarely 6), dehiscing longitudinally; filaments distinct to often monadelphous (occasionally 1 or 2 distinct to base), anthers uniseriate to mostly biseriate; pollen grains globose, binucleate, inaperturate, exine with massive hexagonally arranged processes. *Pistillate flowers* mostly as in the staminate except calyx-lobes more often glandular margined or foliaceous; petals as in the staminate except often larger; disc glands as in the staminate but larger; ovary mostly of 3 carpels but in some taxa reduced to 2 or 1, smooth, glabrous or pubescent; ovules 1 per locule; styles 3 (or often 2 or 1), bifid (sometimes dilated, rarely multifid), variously connate or free, stigma lobes massive or narrow and long. *Fruit* a 1-3 locular capsule, ± fleshy or dry, explosively or often tardily septicidally-loculicidally dehiscent, separating into individual cocci. *Seeds* ellipsoid to spherical, caruncle large and distinct or often reduced or vestigial, gray-brown, black, or yellow in color, solid or variously mottled with brown or black spots or lines; testa crustaceous, starchy endosperm present and copious; embryo spathulate; cotyledons generally broad and palmate-nerved, radicle short, germination phanerocotylar (rarely cryptocotylar). Chromosome number 2n = 22 (rarely 2n = 44).

Below we offer a synoptic key to the infrageneric taxa of *Jatropha,* followed by a checklist of the species. This key is intended to serve as a guide to the groups of species, but cannot be used to unequivocally key each species to the proper section or subsection. The limits between some sections are still not well defined, and discovery of new species may well necessitate changes in circumscription. Many species are still imperfectly known, and further elucidation of their morphological characters will probably suggest shifts in sectional or subsectional assignment.

SYNOPTIC KEY TO THE SUBGENERA, SECTIONS, AND SUBSECTIONS

A. Monoecious. Calyx lobes free and distinct to connate, neither distinctly imbricate nor foliaceous, sometimes with glandular margins. Petals free, coherent, or connate. Stamens 8-10, free or monadelphous. Carpels always 3; fruit trilocular; seeds oblong, with large caruncle. Stipules entire to glandular-dissected. Laticifers mostly articulated and nonarticulated. Trichomes unicellular and multicellular, never verrucose. . . . . Subgenus I. *Jatropha*

  B. Leaves distinctly petiolate, glandular-dentate, often pubescent to villose. Stamens 8 (rarely 6 or 10), monadelphous. Petals free or united.

    C. Flowers campanulate; petals free or basally coherent, often recurved, red to purple or greenish-yellow. South America, Africa, India. . . . . Section I.1. *Jatropha*

      D. Petiole usually with stipitate, sometimes branched, glands; stipules gland-tipped. Inflorescence usually paniculate and lax. Styles free nearly to base; ovary glabrous or pubescent. . . . . Subsection I.1.A. *Adenophorae*

      D. Petiole glandless; stipules filiform, usually not gland-tipped. Inflorescence relatively compact. Styles free or connate; ovary glabrous. . . . . . . Subsection I.1.B. *Pubescentes*

    C. Flowers tubular; petals connate or coherent 1/3 to 3/4 their lengths, lobes not recurved, greenish-yellow to white. Africa.

      E. Coflorescence reduced to a single pistillate flower. Petals connate at least 3/4 their lengths. Petioles 2-3 times as long as leaf-blades; stipules reduced to clustered glands. . . . . Section I.2. *Collenucia*

      E. Coflorescence not evident. Petals connate 1/3 to 1/2 their lengths. Petioles shorter than blades, or suppressed; stipules of 1-3 bifurcate spines. . . . Section I.3. *Spinosae*

  B. Leaves without glandular margins, or if glandular then sessile or subsessile to short-petiolate (petioles mostly less than 2 cm long). Stamens 8-10, free to monadelphous. Petals free.

    F. Leaves glabrous, peltate or subpeltate or else with pinnatifid lobes, palmately veined. Petiole long and relatively stout; stipules distally dissected, rarely glandular. Petals separated by broad sinuses. Stamens 8, filaments free. . . . . Section I.4. *Peltatae*

    F. Leaves glabrous or pubescent, not peltate, lobed or unlobed. Petiole very short to elongated, slender; stipules dissected to base into gland-tipped segments, or obsolete. Petals mostly imbricate. Stamens 8-10, monadelphous.

      G. Stamens 8; calyx lobes entire or denticulate. Petals imbricate or not, greenish-yellow to yellowish-brown. Disc-glands large and conspicuous. . . . . Section I.5. *Tuberosae*

        H. Leaves glabrous or pubescent, entire or lobed, short-petioled or subsessile, margins sometimes glandular-ciliate; stipules usually well-developed, glandular-dissected (rarely reduced). Flowers mostly large (corolla often 1 cm long or more), dish-shaped, petals straight, yellowish-brown. . . . . Subsect. I.5.A. *Tuberosae*

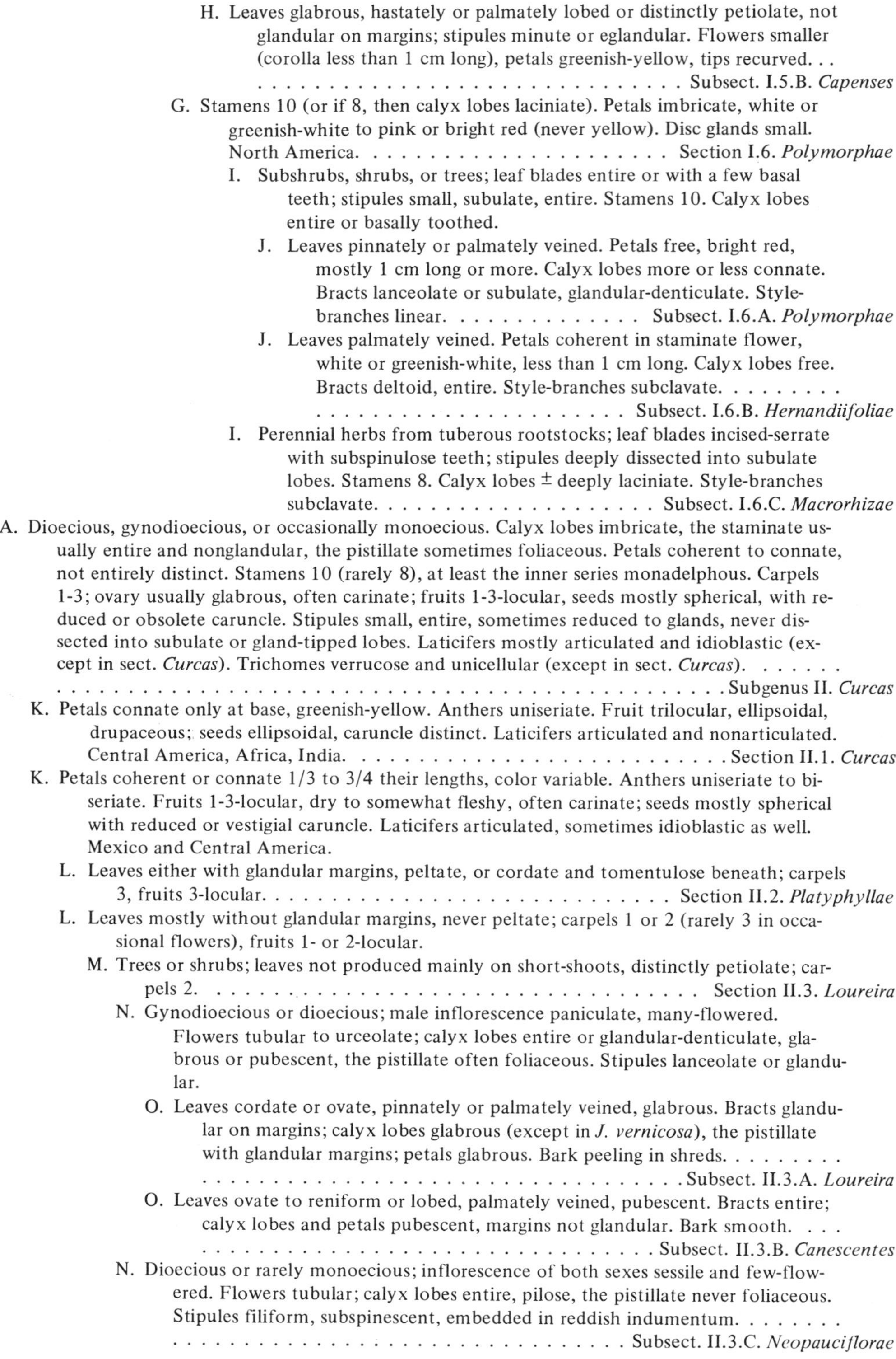

H. Leaves glabrous, hastately or palmately lobed or distinctly petiolate, not
glandular on margins; stipules minute or eglandular. Flowers smaller
(corolla less than 1 cm long), petals greenish-yellow, tips recurved. . .
. . . . Subsect. I.5.B. *Capenses*
G. Stamens 10 (or if 8, then calyx lobes laciniate). Petals imbricate, white or
greenish-white to pink or bright red (never yellow). Disc glands small.
North America. . . . . Section I.6. *Polymorphae*
I. Subshrubs, shrubs, or trees; leaf blades entire or with a few basal
teeth; stipules small, subulate, entire. Stamens 10. Calyx lobes
entire or basally toothed.
J. Leaves pinnately or palmately veined. Petals free, bright red,
mostly 1 cm long or more. Calyx lobes more or less connate.
Bracts lanceolate or subulate, glandular-denticulate. Style-
branches linear. . . . . Subsect. I.6.A. *Polymorphae*
J. Leaves palmately veined. Petals coherent in staminate flower,
white or greenish-white, less than 1 cm long. Calyx lobes free.
Bracts deltoid, entire. Style-branches subclavate. . . . .
. . . . Subsect. I.6.B. *Hernandiifoliae*
I. Perennial herbs from tuberous rootstocks; leaf blades incised-serrate
with subspinulose teeth; stipules deeply dissected into subulate
lobes. Stamens 8. Calyx lobes ± deeply laciniate. Style-branches
subclavate. . . . . Subsect. I.6.C. *Macrorhizae*
A. Dioecious, gynodioecious, or occasionally monoecious. Calyx lobes imbricate, the staminate usually entire and nonglandular, the pistillate sometimes foliaceous. Petals coherent to connate,
not entirely distinct. Stamens 10 (rarely 8), at least the inner series monadelphous. Carpels
1-3; ovary usually glabrous, often carinate; fruits 1-3-locular, seeds mostly spherical, with reduced or obsolete caruncle. Stipules small, entire, sometimes reduced to glands, never dissected into subulate or gland-tipped lobes. Laticifers mostly articulated and idioblastic (except in sect. *Curcas*). Trichomes verrucose and unicellular (except in sect. *Curcas*). . . . . . .
. . . . Subgenus II. *Curcas*
K. Petals connate only at base, greenish-yellow. Anthers uniseriate. Fruit trilocular, ellipsoidal,
drupaceous; seeds ellipsoidal, caruncle distinct. Laticifers articulated and nonarticulated.
Central America, Africa, India. . . . . Section II.1. *Curcas*
K. Petals coherent or connate 1/3 to 3/4 their lengths, color variable. Anthers uniseriate to bi-
seriate. Fruits 1-3-locular, dry to somewhat fleshy, often carinate; seeds mostly spherical
with reduced or vestigial caruncle. Laticifers articulated, sometimes idioblastic as well.
Mexico and Central America.
L. Leaves either with glandular margins, peltate, or cordate and tomentulose beneath; carpels
3, fruits 3-locular. . . . . Section II.2. *Platyphyllae*
L. Leaves mostly without glandular margins, never peltate; carpels 1 or 2 (rarely 3 in occa-
sional flowers), fruits 1- or 2-locular.
M. Trees or shrubs; leaves not produced mainly on short-shoots, distinctly petiolate; car-
pels 2. . . . . Section II.3. *Loureira*
N. Gynodioecious or dioecious; male inflorescence paniculate, many-flowered.
Flowers tubular to urceolate; calyx lobes entire or glandular-denticulate, gla-
brous or pubescent, the pistillate often foliaceous. Stipules lanceolate or glandular.
O. Leaves cordate or ovate, pinnately or palmately veined, glabrous. Bracts glandular on margins; calyx lobes glabrous (except in *J. vernicosa*), the pistillate
with glandular margins; petals glabrous. Bark peeling in shreds. . . . .
. . . . Subsect. II.3.A. *Loureira*
O. Leaves ovate to reniform or lobed, palmately veined, pubescent. Bracts entire;
calyx lobes and petals pubescent, margins not glandular. Bark smooth. . . .
. . . . Subsect. II.3.B. *Canescentes*
N. Dioecious or rarely monoecious; inflorescence of both sexes sessile and few-flow-
ered. Flowers tubular; calyx lobes entire, pilose, the pistillate never foliaceous.
Stipules filiform, subspinescent, embedded in reddish indumentum. . . . .
. . . . Subsect. II.3.C. *Neopauciflorae*

M. Shrubs or rhizomatous subshrubs, most leaves produced on short-shoots. Leaves petiolate or subsessile, entire to deeply lobed. Carpels usually 1 (rarely 2), fruit usually
with a single seed. . . . . .Section II.4. *Mozinna*

SUBGENUS I. *JATROPHA*

Adenoropium Pohl, Pl. Brasil. Icon. Descr. I:12.1827.—*Jatropha* subg. *Adenoropium* Pax in Engler &
Prantl, Natürl. Pflanzenfam. 3(5):75.1890; Pax in Engler, Pflanzenreich IV.147(Heft 42):23.1910; Pax
and Hoffm. in Engler & Prantl, Natürl. Pflanzenfam. ed. 2, 19c:160.1931.

Trees, shrubs, geophytes with subterranean tubers, semi-herbaceous or rarely facultative annuals. Monoecious. Calyx lobes not distinctly imbricate, scarious, subherbaceous or rarely herbaceous; petals distinct
or coherent to connate. Stamens mostly 8(10), monadelphous, uni- or biseriate or free. Styles always 3,
bifid. Fruit trilocular, usually globose, dry, explosively dehiscent. Seed usually oblong and characteristically
prominently carunculate. Laticifers mostly articulated and nonarticulated.

Lectotype species: *Jatropha gossypiifolia* L.

I.1. Section *Jatropha*

Adenoropium Pohl, Pl. Brasil. Icon. Descr. 1:12. t. 9. 1827.—*Jatropha* sect. *Adenorhopium* Griseb. Fl. Brit.
West Ind. 36.1859; McVaugh, Bull. Torr. Bot. Club 72:279.1945 (p.p.).
Jatropha subgen. *Adenoropium* sect. *Glanduliferae* Pax in Engler, Pflanzenreich IV.147(Heft 42):23.1910.

Shrubs, subshrubs, semi-herbaceous, geophytes or rarely facultative annuals. Leaves distinctly 3-5 lobed,
glandular-denticulate on the margins, glabrous or pubescent; petiole 5-10 cm long, with or without solitary
or branched stipitate glands, vascular traces 5 + 2 (7 + 2) medullated, U-shaped; stipules glandular or filiformly dissected with glandular tips. Inflorescences terminal on branches, paniculate, lax or compact, peduncle 5-12 cm long; coflorescence distinguishable; apical dominance absent; bracts oblanceolate or lanceolate with glandular margins. Calyx lobes ovate-lanceolate, acuminate, herbaceous, beset with stipitate glands
at the margins, distinctly imbricate, sparsely or densely pubescent; petals obovate, often recurved (in American taxa), ± connate at the base, yellow, red to purplish. Disc glands distinct and producing copious nectar. Stamens 8, usually uniseriate-monadelphous. Styles 3, bifid, nearly free to distinctly connate; stigma
lobes capitate or horse-shoe shaped, thickened and fleshy; ovary pubescent or glabrous. Capsule relatively
small, 1-1.5 cm long. Seeds small, distinctly carunculate, yellowish or with dark lines or slightly mottled.

Lectotype species: *Jatropha gossypiifolia* L.

This section agrees in general with McVaugh's section *Adenorhopium* but differs in two
respects. First, *J. macrorhiza,* which was doubtfully assigned to this section by McVaugh because of its tubular corolla, dissected stipules, and fimbriate caruncle, has been removed. The
presumably tubular corolla has already been explained to be the result of nyctonastic and/or
thermonastic behavior of the flower. Examination of such features as dissected stipules and
calyx lobes, as well as the imbricate aestivation of the petals, reveals closer affinity to section
Polymorphae, in which it has been placed in the present revision; experimental data (Dehgan,
1976) supports this. Inclusion of African species in subsection *Pubescentes* of section *Jatropha* is in agreement with Pax's treatment. The Indian species, however, show closer affinity
to the American taxa and are included together with them in subsection *Adenophorae.* Recognition of distinct subsections for the American and Indian species based on flower color
or degree of connation of the petals appears not warranted at this time. If in the future, morphological and anatomical characters are recognized that would distinguish these taxa, then
their separation may become necessary.

The sectional name *Jatropha* must replace the well-known *Adenorhopium* in accordance
with the International Code of Botanical Nomenclature (1972), since it includes the type species of the genus.

I.1.A. SUBSECTION *ADENOPHORAE* PAX EX DEHGAN & WEBSTER, SUBSECT. NOV.[1]

Jatropha sect. *Glanduliferae* subsect. *Adenophorae* Pax in Engler, Pflanzenreich IV.147(Heft 42):25.1910 (in clavi; nom. subn.).

Frutices vel herbae, foliis lobatis, petiolis stipulisve stipitato-glandulosis; inflorescentiis paniculatis, laxis; petalis basaliter connatis; stylis fere liberis; ovario pubescenti.

Shrubs, herbaceous subshrubs with or without subterranean tubers, or facultative annuals. Leaves deeply lobed or nearly parted; petiole often with solitary or branched stipitate glands, vascular traces 5 + 2 (rarely 7 + 2 or 9 + 2), medullated, U-shaped, stipules branched or unbranched, gland-tipped. Inflorescence paniculate, lax, 8-12 cm or more long. Petals often recurved, connate at base, greenish-yellow, red, or purplish. Styles free nearly to the base; stigmata capitate, thickened and fleshy; ovary pubescent. Species South American and Indian.

Lectotype species: *Jatropha gossypiifolia* L.

Additional examples: *J. excisa* Griseb., *J. clavuligera* Muell. Arg., *J. katherinae* Pax, *J. elliptica* (Pohl) Muell. Arg., *J. peiranoi* Lourt. & O'Donn., *J. isabellii* Muell. Arg., *J. dissecta* (Chod. & Hassl.) Pax, *J. hippocastanifolia* Croiz., *J. glandulifera* Roxb., *J. tanjorensis* Ellis & Saroja, et al.

I.1.B. SUBSECTION *PUBESCENTES* PAX EX DEHGAN & WEBSTER, SUBSECT. NOV.

Jatropha sect. *Glanduliferae* subsects. *Lobatae, Pubescentes* Pax in Engler, Pflanzenreich IV.147(Heft 42): 25.1910 (in clavi; nom. subn.).

Fruticuli caulibus basaliter plusminusve tumidis, foliis lobatis, petiolis stipulisve eglandulosis; inflorescentiis compactis; petalis basaliter connatis, luteis; stylis liberis vel alte connatis; ovario glabrove pubescenti.

Subshrubs, often with a ± swollen base. Leaves lobed to about 1/2 or less of their length; petiole without stipitate glands; vascular traces 7, medullated, U-shaped; stipules dissected, long and filiform, gland-tipped. Inflorescences corymbose, compact, peduncle 5-8 cm or less long. Petals connate to about 1/3 their length, usually straight, greenish-yellow. Styles free to connate for most of their length, stigmata horseshoe-shaped, thick and fleshy; ovary glabrous or pubescent. Species all African.

Lectotype species: *Jatropha acerifolia* Pax (= *J. velutina* Pax & Hoffm.). Additional examples: *J. brockmanii* Hutchinson, *J. phillipseae* Rendl., *J. spicata* Pax, *J. stuhlmanii* Pax, et al.

I.2. Section *Collenucia* (Chiov.) Chiov.

Collenucia Chiov. Fl. Somala 1:177.1929.–*Jatropha* sect. *Collenucia* (Chiov.) Chiov. Ann. Bot. Roma 18: 323.1930.

Multibranched undershrubs with thick branches and caudex. Leaves small, shortly 3-5 lobed or undulate, often orbicular in outline, rigidly subchartaceous, villose on both surfaces, glandular-dentate at the margin; petiole 2-3 times longer than the blade, villose, not glandular, vascular traces 5, free in a ring; stipules reduced to a mass of persistent glands. Inflorescence lateral, cymose, few-flowered; coflorescence present but reduced to a single pistillate flower; apical dominance absent; bracts oblong-lanceolate, glandular toothed, densely pubescent. Calyx lobes acute, pubescent, margins glandular-stipitate; petals connate in the lower 2/3 to 3/4, forming a definite tubular corolla, greenish-yellow, densely pubescent adaxially, lobes ovate or rounded at the apex, not recurved. Disc gland oblong-obovoid, large and distinct. Stamens 8, filaments connate to near the top, distinctly biseriate. Ovary ovoid-globose, tomentose-pilose; styles free nearly to the base, 3, bifid; stigmata thickened and fleshy. Fruit trilocular, small, less than 1 cm long and wide. Seeds small, yellowish-white, not mottled or lined.

Type species: *Jatropha paradoxa* (Chiov.) Chiov.

---

1. The nomenclature of subsectional names in *Jatropha* is complicated by the fact that Pax (1910) did not provide adequate descriptions. They are mostly *nomina subnuda,* and there is considerable doubt therefore whether they were validly published, although it could be argued that the couplets of the keys to taxa within sections form a sort of description. The final treatment by Pax and Hoffman (1931) does not resolve the ambiguity, as the subsectional names there are still *nomina subnuda.* To resolve this difficulty, we are adopting Pax's names wherever possible, but are providing Latin diagnoses in order to validate them.

This section was originally recognized by Chiovenda as a distinct genus, *Collenucia* (1929), and later (1930) as a section of subgenus *Curcas* (Adans.) Griseb. which included the two species: *J. paradoxa* and *J. marginata.* This was done on the basis of the tubular corolla and reduced inflorescence. It differs, however, from sections *Curcas, Platyphylla, Loureira,* and *Mozinna* (*sensu* Dehgan & Webster) in several major characters (viz. trilocular fruit and 8 stamens, glandular herbage, inflorescence, general growth habit, etc.) and, as indicated earlier, represents a good example of parallel evolution, with respect to connation of the petals, in two widely separated geographical areas. The section is otherwise closely related to section *Jatropha* and represents a more advanced evolutionary stage analogous to section *Mozinna* in subgenus *Curcas.*

I.3. Section *Spinosae* Pax

Jatropha subgen. *Adenoropium* sect. *Spinosae* Pax in Engler, Pflanzenreich IV.147(Heft 42):55.1910.

Arborescent shrubs or subshrubs. Leaves often crowded on short cushion-shaped branchlets in the axils of 2 rigid spiny stipules, lobed, ovate-orbicular, or spathulate-oblanceolate, minutely denticulate at the margin, glabrous on both surfaces or shortly pubescent above and tomentose below; petiole shorter than the blade, glabrous or tomentose; stipular spines branched or distinct, minute to 2 cm long, acute, rigid Cymes terminal or subterminal, short and many-flowered or long and few-flowered, tomentose or glabrous; bracts lanceolate to triangular-ovate, glabrous or pubescent, slightly glandular margined or entire. Calyx lobes ± scarious with few or many glands at the margin in female, and entire or slightly glandular in the male, linear-lanceolate or oblong, acuminate or rounded at the apex, glabrous or villose; disc gland small. Petals entire or clawed, pubescent or glabrous abaxially, glabrous adaxially, greenish-yellow or cream-colored. Stamens 8 (rarely 10), biseriate-monadelphous. Styles slender, connate to the middle, 3, bifid; ovary tomentose or glabrous. Fruit trilocular, small or to more than 2 cm long. Seeds with large, filiformly dissected caruncle.

Type species: *Jatropha spinosa* (Forsk.) Vahl.

Seven species were included in this section by Pax: *J. spinosa, J. ellenbeckii, J. tropaeolifolia, J. ferox, J. rivae, J. fissispina,* and *J. crinita.* Two distinct subsections could reasonably be recognized: one to include *J. ferox* which seems to be an aberrant species (including the long spines) for the section, and for the Old World species of *Jatropha* in general, in having 10 stamens rather than 8. The second subsection would then include the remaining six species which show a much closer relationship to each other in overall morphology than they do to *J. ferox.* In the absence of more living specimens of these taxa, however, it does not seem wise to create two subsections here at this time.

Pax (l.c.) suggested a close similarity between the xeric Mexican taxa of sections *Loureira* and *Mozinna* (*sensu* Dehgan & Webster) and those of section *Spinosae.* However, no spinose species are known from Mexico and the trilocular fruit, 3 bifid styles, glandular herbage, and globose seeds with large caruncle of the African plants are characteristically those of sections *Jatropha* and *Collenucia* (*sensu* Dehgan & Webster). It is, therefore, clear that similarities of the leaves in a few of the species are coincidental and probably the result of adaptation to ecologically similar environments.

I.4. Section *Peltatae* (Pax) Dehgan & Webster, stat. nov.

Jatropha subgen. *Adenoropium* sect. *Glanduliferae* subsect. *Peltatae* Pax in Engler, Pflanzenreich IV.147 (Heft 42):43.1910; subsect. *Multifidae* Pax, op. cit. 40 (nom. subn.).
Jatropha sect. *Macranthae sensu* McVaugh, Torr. Bot. Club Bull. 72:275. 1945 (p.p., excl. typ.).

Abores vel frutices, caulibus plusminusve succulentis, foliis saepe peltatis, petiolis longis, stipulis glandulosis vel dissectis; inflorescentiis corymbosis; petalis liberis, luteis vel rubris; staminibus 8, liberis; stylis liberis; ovario glabro.

Glabrous trees or shrubs with succulent fleshy trunk and branches, or with a caudex, or rarely herbaceous perennials with subterranean tubers. Leaves often peltate or subpeltate, shallowly 3-5 lobed or deeply divided; petiole long and stout, with 11 (rarely 9 or 7) vascular traces in a medullated cyinder; stipules stiff-glandular or filiformly dissected. Inflorescence pseudoterminal, corymbose; peduncle 9-25 cm long; coflorescence absent, apical dominance not evident; bracts small, entire, non-glandular. Calyx lobes scarious, ovate-triangular, short and blunt, entire and glabrous in flowers of both sexes; petals 1-1.5 cm long, distinct, glabrous, obovate, yellow, red to scarlet. Disc glands deeply 5-lobed in female, or completely free in male, ovoid. Stamens 8, free; exinous knobs of the pollen grain without the typical striations. Ovary glabrous, styles 3, bifid, distinct; stigmata short and capitate. Fruit dry, trilocular. Seeds ellipsoidal, with small scar and large fimbriate caruncle (except in *J. multifida*).

Type species: *Jatropha peltata* H.B.K. (= *J. humboldtiana* McVaugh).

Exclusion of *J. macrantha* as the type species of section *Macranthae* Pax necessitates the change of name to section *Peltatae,* which is basically the same in circumscription as that of section *Macranthae sensu* McVaugh. The section as indicated in the historical account was a homogeneous group as organized by McVaugh, except for the *J. macrantha* group which has 10 stamens (rather than 8, as is characteristic for the section). The free stamens set section *Peltatae* apart from all other groups within *Jatropha.*

The African species which were assigned to this section by Pax (1.c.), as for example *J. spicata* Pax, have been removed. They appear from descriptions to be more closely allied to subsection *Pubescentes* of section *Jatropha.* Since Pax's descriptions are not explicit and no living plants of these taxa have been available for study to ascertain whether the stamens are free in the African species, definite assignment will have to await the future. The section is otherwise South American with a tuberous relict species, *J. cathartica* Terán & Berl., in the northern Chihuahuan desert of southern Texas. Other species can be exemplified by *J. multifida* L., *J. augustii* Pax & Hoffm., *J. hieronymii* Kuntze, *J. podagrica* Hook., et al.

I.5. Section Tuberosae Pax, emend.

Jatropha subgen. *Adenoropium* sect. *Tuberosae* Pax in Engler, Pflanzenreich IV.147(Heft 42):59.1910; incl.
   subsect. *Latifoliae* Pax and subsect. *Macrorhizae* Pax (p.p.).

Small shrubs or subshrubs with a swollen base and/or subterranean tubers. Leaves variable in shape; entire or serrate with small, glandular teeth; short petiolate or subsessile with 7 vascular traces in a ring or as free traces; stipules often glandular or setaceous and without glandular tips. Inflorescence terminal; cymes dense and few-flowered or lax and with many flowers; coflorescence present and distinct or present but indistinguishable; bracts lanceolate, glandular toothed or often entire, 1-4 cm long. Calyx scarious, lobes entire, slightly toothed or glandular; petals free with a broad sinus or imbricate, spathulate-lanceolate or oblanceolate-obovate, greenish-yellow to yellow-brown. Disc glands 5, free, large and conspicuous. Stamens 8, mostly biseriate-monadelphous. Ovary glabrous; styles 3, bifid, free or connate near the base, stigmas massive. Fruit trilocular, dry, slightly lobed. Seeds yellow or light brown with a relatively large fimbriate caruncle.

Type species: *Jatropha tuberosa* Pax

The diversity of this section is immediately apparent in the above description, and logic dictates division of the taxon into two different subsections. *Capenses,* one of the two subsections recognized below, includes only two South African species—*J. capensis* (L.f.) Sond. and *J. variifolia* Pax—that are seemingly more closely related to section *Curcas* than they are to any other section in the genus on the basis of floral structure. Vegetatively, however, they differ considerably from section *Curcas* and also among themselves. Certain morphological features as well as growth habit are very similar to subsection *Tuberosae,* hence their inclusion in the same section.

Of the five African species that were assigned by Pax to his section *Polymorphae,* three are referable to subsection *Capenses. Jatropha unicostata* Balf. f. is similar in floral structure to members of section *Tuberosae* Pax (*sensu stricto*), e.g. *J. gallabatensis* Schweinf., *J. woodii* Kuntze, *J. lagarinthoides* Sond., et al., and therefore, is assigned to subsection *Tuberosae* Pax.

Section *Tuberosae* is, therefore, exclusively African (except for the dubiously placed Indian species *J. nana* Dalz. & Gibson and *J. heynei* Balakr.). The section is perhaps best represented in the Somali Republic, indicating adaptation to severe xeric conditions by formation of subterranean tubers.

I.5.A. SUBSECTION *TUBEROSAE* DEHGAN & WEBSTER, SUBSECT. NOV.

Frutices vel fruticuli, saepe tuberosis, foliis plerumque elobatis, petiolis brevibus, stipulis glandulosis vel setaceis; inflorescentiis terminalibus cymosis, laxis; petalis liberis, flavidis.

Shrubs or subshrubs, often with subterranean tubers. Leaves ovate to ovate-lanceolate, unlobed to palmately or rarely pinnately lobed, margins entire or often denticulate-glandular; petiole short or subsessile; stipules obsolete to short-glandular or setaceous, sometimes gland-tipped. Inflorescence terminal, cymes lax and many flowered; coflorescence distinct; flowers large, campanulate. Calyx lobes entire, slightly toothed or glandular; petals free with a broad sinus, oblong-obovate, yellowish-brown.

Type species: *Jatropha tuberosa* Pax

Additional examples: *J. gallabatensis* Schweinf., *J. natalensis* Muell. Arg., *J. zeyheri* Sond., et al.

I.5.B. SUBSECTION *CAPENSES* DEHGAN & WEBSTER, SUBSECT. NOV.

Frutices vel fruticuli, foliis hastatis vel palmatilobis, petiolis brevibus; floribus pusillis campanulatis, petalis liberis imbricatis recurvatis spathulatis luteo-viridibus.

Shrubs or subshrubs with a caudex. Leaves oblong-lanceolate, ovate or oblong, hastately or palmately lobed, margins entire or sparsely toothed but not glandular; petiole short but definitely not subsessile; stipules small, gland-like, or subulate and eglandular. Inflorescence terminal, many-flowered, coflorescence not distinct; flowers relatively small and ± tubular. Calyx lobes entire; petals free but with imbricate aestivation, spathulate-lanceolate, greenish-yellow.

Type species: *Jatropha capensis* (L.f.) Sond.

Additional examples: *J. variifolia* Pax, *J. glauca* Vahl, *J. prunifolia* Pax.

I.6. Section *Polymorphae* Pax, emend.

Jatropha sect. *Polymorphae* Pax in Engler, Pflanzenreich IV. 147 (Heft 42):48. 1910; sect. *Macranthae* Pax, op. cit. 46 (p.p.).

Small trees, multibranched shrubs, succulent subshrubs, or herbaceous perennials with subterranean tubers. Leaves variable, pinnately or palmately veined, often with a few glandular subspinulose teeth at the margins, especially near the base, glabrous or nearly so; long-petiolate, vascular traces 7 in a cylinder with secondary growth or free and without secondary growth; stipules narrow, subulate, setaceous, but not glandular (or rarely so). Inflorescence terminal or subterminal, without distinct coflorescence; peduncle 6-12 cm long; apical dominance absent. Bracts lanceolate, entire to lobed or divided. Calyx lobes scarious or subherbaceous, entire or obscurely denticulate, sometimes basally connate; petals imbricate, elliptic-obovate, greenish-white, pink, or bright red. Disc glands five, free, small and inconspicuous. Stamens 8-10, biseriate-monadelphous. Ovary glaucous, styles 3, bifid, free or partially connate, filiform and long or massive and short. Fruit trilocular, dry, explosively dehiscent. Seeds oblong with reflexed and lacerate or fimbriate caruncle, light with dark brown or purplish specks.

Lectotype species: *Jatropha integerrima* Jacq.

Delimitation of section *Polymorphae* is here broadened to include all of the West Indian species of *Jatropha,* including *J. hernandiifolia* and *J. divaricata.* These two species had been assigned to subgenus *Curcas* in all of the past treatments of the genus due to the presumed basal connation of the petals and to flower color. Several lines of evidence, however, indicate that the two species have certain characters of section *Peltatae* (viz. characters of styles, and subclavate stigmata, epidermal morphology, etc.), and others of the section *Polymorphae* (viz. 10 biseriate-monadelphous stamens, imbrication rather than connation of the petals, arrangement of petiolar traces with secondary growth, unmistakable similarity in fruit and seeds, etc.). Their inclusion here as a distinct subsection is, therefore, warranted on the basis of gross morphology and of geographical affinity. *Jatropha macrantha* from Peru and a few other South American taxa are also included here in *Polymorphae* because of striking similarity of reproductive structures (including the 10 stamens). Closer examination of the vegetative morphology indicates probable reduction in the lobes of the leaves and loss of succulence which has probably taken place during evolution in the more mesic environment of the West Indian Islands. The succulent character and deciduous habit of *J. macrantha* and its related taxa are probably ancestral and the non-succulent, multibranched, evergreen growth habit is secondarily derived.

Jatropha macrorhiza, with imbricate aestivation of the petals and several other morphological features discussed earlier, also displays secondary adaptation to a more xeric habitat. It is considered to be a relict of an older 'polymorphian' stock which became widespread in Mexico after its introduction there. This point has been discussed elsewhere (Dehgan, 1976) and further elaboration here does not seem necessary. The recognition of the monotypic subsection *Macrorhizae* is made primarily on the basis of its growth habit and leaf morphology.

I.6.A. SUBSECTION *POLYMORPHAE* (PAX) DEHGAN & WEBSTER, STAT. NOV./Based on *Jatropha* sect. *Polymorphae* Pax, Supra.

Jatropha sect. *Macranthae, sensu* McVaugh, Torr. Bull. Bot. Club 72:275.1945 (p.p.).

Large shrubs or succulent subshrubs. Leaf blades ovate to elliptic, lobed or lanceolate, acuminate, palmately or pinnately veined, often with sclerophyllous vein endings; entire or with a few glandular or subspinulose teeth near the base; petioles long with 7 vascular traces in a cylinder and with secondary growth; stipules small, subulate and entire. Inflorescence corymbose, pseudoaxillary; coflorescence not distinct; bracts long, ovate-lanceolate, toothed or subulate. Calyx campanulate, lobes entire or toothed; petals imbricate, bright red, hirsutulous at base within. Stamens 10, biseriate-monadelphous. Stigmata long and linear.

Lectotype species: *Jatropha integerrima* Jacq.
Additional examples: *J. angustifolia* Griseb., *J. tupifolia* Griseb., *J. paxii* Croiz., *J. macrantha* Muell. Arg., et al.

I.6.B. SUBSECTION *HERNANDIIFOLIAE* DEHGAN & WEBSTER, SUBSECT. NOV.

Arbores vel frutices, foliis palmatinerviis, petiolatis; bracteis parvis acuminatis; laciniis calycis integris; petalis liberis vel cohaerentibus, cloroleucis; stigmatibus brevibus, subclavatis.

Large shrubs or small trees. Leaf blades ovate to elliptic or shallowly 3-lobed, nearly peltate or rounded at the base, palmately veined, entire without teeth or glands, glabrous or nearly so; petiole long with 7 vascular traces in a cylinder and with secondary growth; stipules subulate, early deciduous. Inflorescence corymbose, few-flowered, pseudoaxillary; coflorescence absent; bracts small, deltoid, acuminate, entire. Calyx campanulate, lobes entire; petals imbricate in female, slightly coherent in male, greenish-white. Stamens 10, biseriate-monadelphous. Stigmata subclavate.

Type species: *Jatropha hernandiifolia* Vent.
Additional example: *J. divaricata* Sw.

I.6.C. SUBSECTION *MACRORHIZAE* PAX EX DEHGAN & WEBSTER, SUBSECT. NOV.

Jatropha sect. *Tuberosae* subsect. *Macrorhizae* Pax in Engler, Pflanzenreich IV.147 (Heft 42):65.1910 (nom. subn.; p.p., as to type only).

Herbae tuberosae, foliis profunde lobatis, marginibus serratis, petiolis longis, stipulis dissecto-lanciniatis; inflorescentiis pseudo-axillaribus; lobis calycis lanciniato-serratis; petalis liberis, rubellis vel rubris.

Perennial herbs with subterranean tubers. Leaf blades deeply 3-5 (7) lobed, margins incised-serrate with subspinulose teeth. Petiole long, with 7 free traces and without secondary growth; stipules deeply dissected into 3-5 elongate, subulate, setaceous lobes. Inflorescence cymose, terminal; coflorescence indistinguishable; bracts lanceolate-subulate. Calyx lobes connate below, ± laciniate-toothed; petals free, whitish-pink, pink to red, glabrous. Stamens 8, biseriate-monadelphous. Stigmata subclavate.

Type species: *Jatropha macrorhiza* Benth.

As here construed, subsection *Macrorhizae* is monotypic and entirely American; African species included here by Pax should mostly be referred to subsection *Tuberosae*.

SUBGENUS II. *CURCAS* (ADANS.) PAX

Curcas Adans. Fam. Pl. 2:356.1763.—*Jatropha* subgen. *Curcas* Pax in Engler & Prantl, Pflanzenfam. 3(5): 74.1890; Pax in Engler, Pflanzenreich IV.147(Heft 42):74.1910; Pax & Hoffm. in Engler & Prantl, Pflanzenfam. ed. 2, 19c:162.1931.

Jatropha sect. *Mozinna* (Ortega) Pax, *sensu* McVaugh, Torr. Bull. Bot. Club 72:282.1945.

Type species: *Jatropha curcas* L.

Trees, shrubs, or rhizomatous subshrubs. Monoecious, gynodioecious, or dioecious. Calyx lobes imbricate, usually herbaceous and the pistillate often foliaceous. Petals coherent to increasingly connate but never distinct. Stamens always 10, monadelphous, uni- or biseriate but never free. Ovary glabrous, often carinate; styles 3, 2, or 1, bifid. Fruit tri-, bi-, or unilocular, ellipsoidal or distinctly lobed, drupaceous or capsular, ± tardily dehiscent. Seeds mostly spherical with reduced or vestigial caruncle. Laticifers mostly articulated and idioblastic (except in section *Curcas*).

II.1. Section *Curcas* (Adans.) Griseb.

Curcas Adans. Fam. Pl. 2:356.1763.—*Jatropha* sect. *Curcas* Griseb. Fl. Brit. West Ind. Isl. 36.1859.—*Curcas* sect *Eucurcas* Baill. Étud. Gen. Euphorb. 314.1858.—*Jatropha* sect. *Curcas* subsect. *Eucurcas* Muell. Arg. in D.C. Prodr. 15(2):1080.1866.—*Jatropha* sect. *Mozinna* subsect. *Eucurcas* McVaugh, Bull. Torr. Bot. Club 72:283.1945; Wilbur, J. Elisha Mitch. Sci. Soc. 70:92.1954.

Castiglionia Ruiz et Pavon, Fl. Peruv. 139.1794.—*Jatropha* subgen. *Curcas* sect. *Castiglionia* (Ruiz et Pavon) Pax in Engler, Pflanzenreich IV.147(Heft 42):76.1910.

Small trees or large shrubs; bark smooth. Monoecious, dioecious, or rarely some flowers bisexual (in *J. curcas*). Leaves ovate, shallowly (3) 5-7 lobed, palmatinerved, cordate at the base, sparsely puberulent to densely pubescent, sparsely glandular at the margin; long petiolate, petiolar traces 9 in a medullated cylinder; stipules small or nearly obsolete. Inflorescence subterminal with a distinct coflorescence; peduncle 3-12 cm long, axis tomentulose to densely pubescent; bracts lanceolate, entire, tomentose to pubescent. Calyx lobes oblong to lanceolate, entire, ± foliaceous in female; petals obovate, connate at the base, corolla tube usually longer than the lobes. Disc segments 5, ellipsoid, large. Stamens 8-10, uniseriate-monadelphous. Styles 3, bifid; stigmata massive; ovary glabrous. Fruit ellipsoidal, large, 2-3 cm long and broad, ± fleshy. Seeds large, flat, 2 to 2.5 cm long, ellipsoid, brown-black, caruncle appressed and small.

Type species: *Jatropha curcas* L.

Pax (l.c.) in his treatment of this group (as section *Castiglionia*) recognized six species including *J. curcas,* the origin of which cannot be determined with certainty. Two species, *J. yucatanensis* and *J. pseudocurcas,* are American; two, *J. afrocurcas* and *J. macrophylla,* are east African; and the sixth, *J. wightiana,* is Indian. McVaugh (l.c.) considered *J. yucatanensis* to be a synonym of *J. curcas,* but added *J. andrieuxii* to subsection *Eucurcas* of section *Mozinna.* Wilbur (1954) added *J. hintonii* and *J. bartlettii* to the list to bring the total to eight species; and finally *J. mcvaughii* has recently been described from Mexico (Dehgan and Webster, 1978).

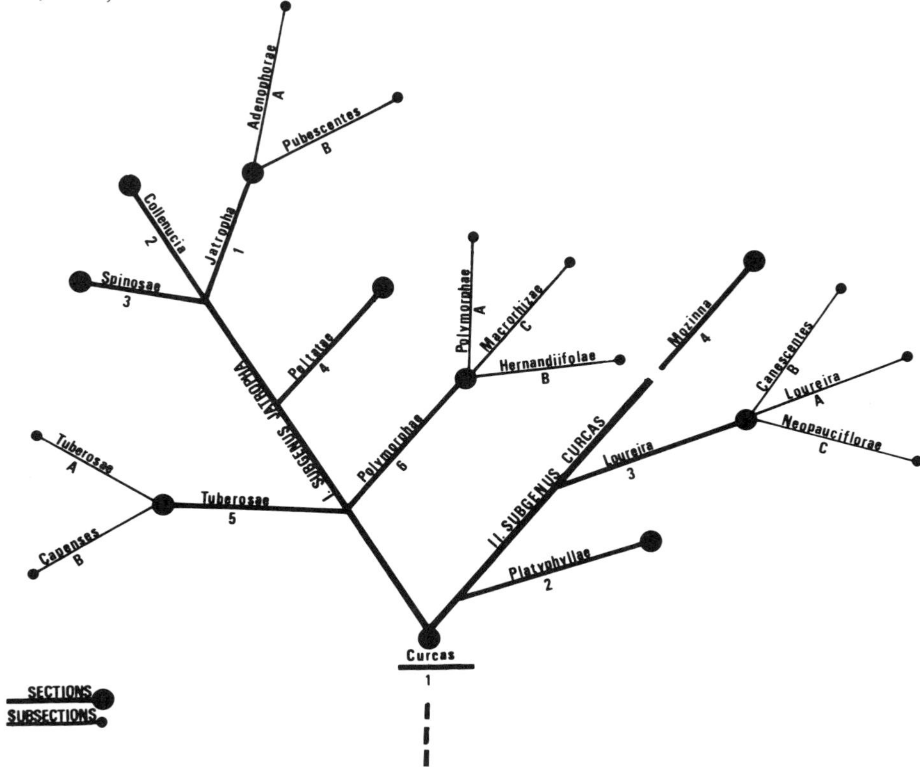

Figure 4. Phylogenetic diagram of the genus *Jatropha* L.

This section has been placed at the base of the phylogenetic diagram (Fig. 4) because it is considered ancestral by having occasional bisexual flowers, distinct coflorescence, large palmately lobed leaves, and arborescent growth habit, as well as a large number of morphological and anatomical characters (including the presence of both articulated and nonarticulated laticifers) that it shares with sections of subgenus *Jatropha*.

II.2. Section *Platyphyllae* Dehgan & Webster, sect. nov.

Arbores, fructices, vel fruticuli; foliis peltatis vel cordatis, 3-5 lobatis, petiolatis; laciniis calycis magnis, ♀ foliaceis; corollis gamophyllis, subglobosis vel tubulosis; stylis 3, bifidis; capsulis 3-lobatis carinatis; seminibus globosis, carunculo exiguo.

Trees, shrubs, or subshrubs; bark smooth or fissured. Leaves ovate-orbicular, peltate or broadly cordate at base, 3-5 lobed, margins entire or with stipitate glands, long-petiolate, petiolar traces 8-9 in a medullated cylinder; stipules small or obsolete, not glandular. Inflorescence subterminal; coflorescence not distinct; monoecious or often dioecious; peduncle 3-8 cm long, axis glabrous; bracts entire or denticulate, glabrous or tomentulose. Calyx lobes glabrous, ovate to obtuse, entire, the pistillate sometimes foliaceous, 2-4 cm or more long in the female, somewhat shorter in the male; petals lanceolate-elliptic, straight, considerably larger in the pistillate flower; corolla tubular to urceolate, petals connate to 1/3 of their length or more, glabrous or hirsutulous within, greenish-white to white. Disc glands large, to 2 mm, distinct in male, free to ± connate in female. Stamens 10, biseriate-monadelphous. Ovary glabrous, styles 3, bifid; stigmata massive. Fruit trilocular, distinctly 3-lobed; lobes carinate. Seed spherical to ellipsoidal, caruncle rather small to minute, brown with yellow stripes or mottles.

Type species: *Jatropha platyphylla* Muell. Arg.

Included in this section of five species, in addition to *J. platyphylla,* are *J. alamanii* Muell. Arg.; *J. ciliata* Sessé; a newly described species from Baja California (*J. moranii* Dehgan & Webster, 1978); and one from Costa Rica (*J. costaricensis* Webster & Poveda, 1978). The section is characterized by campanulate pistillate flowers, tubular to urceolate staminate flowers, and distinctly trilobed carinate fruit.

II.3. Section *Loureira* (Cav.) Muell. Arg. ex Pax

Loureira Cav. Icon. Descr. Pl. 5:17.1799.–*Jatropha* sect. *Loureira* (Cav.) Muell. Arg. ex Pax, Pflanzenreich 147(Heft 42):74.1910.

Shrubs or small trees; bark wrinkled or peeling. Gynodioecious or dioecious (rarely monoecious). Leaves cordate, spathulate, ovate-lanceolate or palmately lobed; margins entire or with stalked or subsessile glands, glaucous to densely pubescent; petiole short or long with 7 free vascular traces; stipules small, narrowly lanceolate or often glandular. Inflorescence terminal and closely paniculate or lateral with subsessile, compact paracladia; coflorescence indistinguishable if present; bracts lanceolate, entire and pubescent or glandular margined and glabrous. Flowers urceolate, subglobose or ± tubular; pistillate flowers larger than staminate and much fewer in number. Calyx lobes glabrous or pubescent, entire or glandular margined, the pistillate sometimes foliaceous. Petals connate 1/2 to 2/3 of their length, often recurved, creamy-white, greenish-yellow, or various shades of red, glabrous adaxially but often puberulent to pubescent abaxially. Disc glands 5, large and distinct. Stamens 10, biseriate-monadelphous. Styles 2 (rarely 3), bifid (or rarely parted); ovary glabrous. Fruit bilocular (rarely trilocular or unilocular by abortion) and distinctly lobed, tardily dehiscent. Seeds subglobose or spherical; caruncle minute or vestigial and represented by a scar.

Type species: *Loureira glandulifera* Cav. [=*Jatropha cordata* (Ortega) Muell. Arg.]

The vegetative morphology of this section is probably the most diverse in the genus. On the basis of reproductive structure, however, it is a coherent group in which three subsections are recognizable. One newly described species related to the *J. cinerea* complex (*J. giffordiana* Dehgan & Webster) from Cabo San Lucas in Baja California, is gynodioecious, has globose flowers, and often with trilocular fruit. The remainder of taxa in the section show a reduction from 3 to 2 bifid styles, and from 3 to 2-locular fruit. The species are widespread in Mexico from Oaxaca to Sonora and the Arizona border.

II.3.A. SUBSECTION *LOUREIRA* (CAV.) MUELL. ARG.

Loureira Cav. Icon. Descr. Pl. 5:17.1799.–*Jatropha* sect. *Curcas* subsect. *Loureira* (Cav.) Muell. Arg. in D.C. Prodr. 15(2):1076.1866.

Small trees or large shrubs; bark peeling in large sheets, yellow, the stem otherwise glossy green (rarely pubescent). Gynodioecious or dioecious. Leaves cordate or ovate-lanceolate, acuminate, pinnately or palmately veined, glabrous, margins partly or entirely glandular; stipules of one to several sessile or stipitate glands; petiole long and narrow or short and stout, glabrous. Staminate inflorescence paniculate, lax, many-flowered, often with a solitary, large, pistillate flower at the base. Pistillate inflorescence nearly subsessile, with 1-4 flowers; flowers generally subglobose; bracts ovate-lanceolate and glandular margined in inflorescence of both sexes. Calyx lobes entire and glaucous in the staminate, but glandular margined and often foliaceous in the pistillate flowers; petals glabrous. Fruit bilocular and distinctly bilobed. Seeds globose.

Type species: *Loureira glandulifera* Cav. (= *Mozinna cordata* Ortega = *Jatropha cordata* Muell. Arg.) Additional examples: *J. ortegae* Standley, *J. standleyi* Steyerm., *J. riojae* Miranda, *J. vernicosa* T. S. Brandegee, et al.

II.3.B. SUBSECTION *CANESCENTES* PAX EX DEHGAN & WEBSTER, SUBSECT. NOV.

Jatropha subgen. *Curcas* sect. *Mozinna* subsect. *Canescentes* Pax in Engler, Pflanzenreich IV.147(Heft 42): 18.1910 (nom. subn.).

Frutices, foliis integris vel leviter lobatis, palmatinerviis, petiolis elongatis, stipulis exiguis; inflorescentiis dioicis; floribus urceolatis; calycis lobis integris, dense pubescentibus, saepe foliaceis; stylis 2, bifidis; capsulis lobatis, bilocularibus.

Shrubs; bark smooth, gray or brown, and pilose. Dioecious (or rarely gynodioecious). Leaves relatively broad, ovate, oblong or reniform, cuneate or rounded at the base, often shallowly lobed, entire or glandular margined when young; palmately veined ± pubescent on both surfaces; long petiolate; stipules small, narrowly lanceolate or absent, often appearing on short shoots. Staminate inflorescence many-flowered, lax to compact, and nearly subsessile; pistillate inflorescence few-flowered, nearly subsessile; flowers urceolate or subglobose; bracts small, ovate-lanceolate, entire, densely pubescent. Calyx lobes pubescent in both sexes, often foliaceous in the female; petal usually pubescent abaxially, often recurved. Fruit bilocular (or rarely trilocular, or unilocular by abortion), distinctly lobed. Seeds spherical with vestigial caruncle.

Type species: *Jatropha canescens* Benth. ex Muell. Arg.

Additional examples: *J. cinerea*—a hybridizing complex which requires clarification; *J. giffordiana* Dehgan & Webster.

II.3.C. SUBSECTION *NEOPAUCIFLORAE* DEHGAN & WEBSTER, SUBSECT. NOV.

Frutices, foliis ovato-cuneatis, palmatinerviis; petiolis brevibus; inflorescentiis dioicis vel monoecis, in ramulis abbreviatis; floribus pusillis tubulosis; calycis lobis integris, dense pubescentibus, non foliaceis; stylis 2, bifidis; capsulis lobatis, bilocularibus.

Shrubs; bark wrinkled, grayish-brown. Dioecious (rarely monecious). Leaves ovate-cuneate, pilose or densely pubescent on one or more often both surfaces, palmately veined, entire; petiole short or stout, pilose or densely pubescent; stipules dissected into many long-filiform segments, persistent and subspinose, and often covered with red or brown hairs. Inflorescence few-flowered in both sexes, never foliaceous; flowers ± tubular; bracts small, lanceolate, pilose or densely pubescent. Calyx lobes entire, eglandular, pubescent, not foliaceous; petals pubescent. Fruit bilocular and distinctly bilobed. Seeds spherical with vestigial caruncle (or rarely ellipsoid with large caruncle?).

Type species: *Jatropha neopauciflora* Pax

Additional examples: *J. sympetala* Standley & Blake and *J. fremontioides* Standley.

II. 4. Section *Mozinna* (Ortega) Pax

Mozinna Ortega, Nov. Rar. Pl. Hort. Bot. Matr. 8:104.1799.—*Curcas* sect. *Mozinna* (Ortega) Baill. Étud.
  Gen. Euphorb. 315.1858.—*Jatropha* subsect. *Mozinna* (Ortega) Muell. Arg. in DC. Prodr. 15(2):1081.
  1866.—*Jatropha* sect. *Mozinna* (Ortega) Pax in Engler & Prantl, Pflanzenfam. 3(5):75.1890.
Jatropha sect. *Mozinna* subsect. *Brachyblastae* Pax in Engler, Pflanzenreich IV. 147(Heft 42):81.1910.
Zimapania Engler & Pax in Engler & Prantl, Pflanzenfam. ed. 1, 3(5):119. 1891.

Shrubs or rhizomatous subshrubs with thickened stem, smooth bark, and characteristic short shoots. Leaves irregularly crenulate, cuneate, or lobed on the long shots, spathulate on the short ones; long petiolate or subsessile with 5(3) free traces; stipules eglandular and early deciduous. Inflorescence compact, shortly pedunculate, terminal in the male or consisting of 1-2(3) subsessile flowers in the leaf axils or at the apex of the short shoots in the female; bracts often 1-2 cm long, ovate-lanceolate and entire. Calyx lobes elliptic, blunt, entire, rarely foliaceous in female; corolla tubular; petals nearly completely connate, often recurved, glabrous without, hirsutulous within near base, greenish-white to pinkish-red in color. Disc glands 5, large and distinct. Stamens 10, biseriate-monadelphous. Style 1, bifid, short; stigma massive; ovary glabrous. Fruit usually unilocular and tardily dehiscent. Seed solitary, spherical; caruncle vestigial and represented by a scar.

Type species: *Mozinna spathulata* Ortega (= *Jatropha dioica* Sessé)

The section in its present circumscription includes three species: *J. dioica* Sessé, *J. cuneata* Wiggins & Rollins, and *J. cardiophylla* (Benth.) Muell. Arg. in the Sonoran and Chihuahuan deserts. The low number of petiolar traces (5 or 3), the single style, and the tardily dehiscent unilocular fruit, in association with the rhizomatous growth habit and tetraploid chromosome number, indicate its distinction from the other sections as the most advanced stage of the evolutionary reduction series which begins with section *Platyphyllae*.

CHECKLIST OF SPECIES

In the following enumeration are listed all of the specific and subspecific names (except below the rank of variety) which have been applied to taxa now classified in *Jatropha.* Names originally published in *Jatropha* which have been transferred to *Cnidoscolus* are not accounted for here, but have been indexed by McVaugh (1944, 1945b). Names which have been transferred to other genera are listed by Mueller (1866), Pax (1910), and Rogers & Appan (1973, for *Manihot*).

Names given in boldface represent species accepted by Pax or by more recent workers, while italicized names are synonymous or unacceptable. We have tried to avoid making taxonomic judgements by listing the names generally accepted in the literature, even though it seems probable that some of these will eventually be reduced to synonymy; for example, Allem (1977) has recently proposed to combine a number of South American species with *J. isabellii.* We have attempted to refer each species to its proper section or subsection, but the position of ones known to us only from descriptions are often dubious (and then are indicated with a question mark). Literature citations have been abbreviated (except where this might create ambiguity); names such as var. *genuina* or *typica* are omitted.

Type localities have been given for each taxon, and where the distribution of the species is known to extend into another major political unit, this is indicated in brackets. The range of some taxa, however, may be considerably more extensive than our citation would suggest.

acerifolia Salisb. (1796): see **curcas**

acerifolia Pax (1894): see **velutina**

aceroides (Pax & Hoffm.) Hutchinson, Fl. Trop. Afr. 6(1): 789. 1913; Andrews, Fl. Pl. Anglo-Egypt. Sudan 2: 83. 1952. (I. 1. B.)
>Syn.: *J. lobata* ssp. *aceroides* Pax & Hoffm. Pflanzenreich 147. (Heft 42): 34. 1910.
>Type locality: Ataba, Nubia, Sudan.

aculeata F. G. Dietr.: see **spinosa**

acuminata Desr.: see **integerrima**

aethiopica Muell. Arg. Flora 47: 485. 1864; in DC. Prodr. 15(2): 1088. 1866; Pax, Pflanzenreich 147(Heft 42): 70. 1910; Hutchinson, Fl. Trop. Afr. 6(1): 775. 1913; Andrews, Fl. Pl. Anglo-Egypt. Sudan 2: 85. 1952. (I. 5. A.)
>Syn.: *J. sabdariffa* Schweinf. Beitr. Fl. Aethiop. 1: 37. 1867.
>Type locality: Ethiopia.

afrocurcas Pax, Bot. Jahrb. Syst. 43: 83. 1899; Pflanzenreich 147 (Heft 42): 79. 1910. (II. 1; = **curcas** ?)
>Type locality: Sseke, Tanzania.

alamanii Muell. Arg. Linnaea 34: 208. 1865; in DC. Prodr. 15(2):1077. 1866; Pax, Pflanzenreich 147 (Heft 42): 76, fig. 29. 1910; Standley, Contr. U. S. Nat. Herb. 23: 639. 1923; McVaugh, Bull. Torrey Bot. Club 72: 289. 1945. (II. 2.)
>Type locality: Tehuantepec, Oaxaca, Mexico.

andrieuxii Muell. Arg. Linnaea 34: 208. 1865; in DC. Prodr. 15(2): 1082. 1866; Pax, Pflanzenreich 147
(Heft 42): 48. 1910; Standley, Contr. U. S. Nat. Herb. 23: 636. 1923; McVaugh, Bull. Torrey Bot.
Club 72: 284. 1945; Wilbur, J. Elisha Mitchell Sci. Soc. 70: 98. 1954. (II. 1.)
Type locality: Puebla to Oaxaca, Mexico.

angustifolia Griseb. Abh. Konigl. Ges. Wiss. Göttingen 12: 171. 1865; Muell. Arg. in DC. Prodr. 15(2):
1093. 1866; Pax, Pflanzenreich 147 (Heft 42): 52. 1910; McVaugh, Bull. Torrey Bot. Club 72: 274.
1945; Alain, Fl. Cuba 3: 77. 1953. (I. 6. A.)
var. **angustifolia**
Syn.: *Adenoropium angustifolium* (Griseb.) Arthur, Torreya 21: 11. 1921.
Type locality: western Cuba.
var. **spathulata** Muell. Arg. in DC. Prodr. 15(2): 1093. 1866.
Syn.: *J. glauca* Griseb. Abh. Konigl. Ges. Wiss. Göttingen 12: 170. 1865 (non Vahl 1790).—*J. angusti-*
folia var. *glauca* (Griseb.) Pax, Pflanzenreich 147(Heft 42): 52. 1910.
Type locality: western Cuba.

antisyphilitica Speg.: see **isabellii** var. **antisyphilitica**

arguta Chiov. Fl. Somala 1: 307, t.35 fig. 3. 1929. (I. 3.)
Type locality: between Tigieglo and Bur Cal, Somali Republic.

arizonica I. M. Johnston: see **macrorhiza** var. **septemfida**

aspleniifolia Pax, Bot. Jahrb. Syst. 19: 108. 1894; Pflanzenreich 147 (Heft 42): 83. 1910; Hutchinson, Fl.
Trop. Afr. 6(1): 780. 1913. (I. 5?)
Type locality: foothills of the Ahl Mountains, Somali Republic.

atacorensis Cheval. Bull. Soc. Bot. France 58, Mém. 8: 206. 1911; Pax & Hoffm. Pflanzenreich 147, Addit.
V(Heft 63): 399. 1914; Hutch. & Dalz. Fl. W. Trop. Afr. ed. 2, 1:397.1958. (I. 5. A?)
Type locality: Atacora Mountains, Dahomey.

augustii Pax & Hoffm. Pflanzenreich 147, Addit. VII (Heft 85): 191. 1924; McVaugh, Bull. Torrey Bot.
Club 72: 277. 1945; Macbride, Field Mus. Nat. Hist., Bot. Ser. 13(3A): 159. 1951. (I. 4.)
Syn.: *J. ciliata* Muell. Arg. Linnaea 34: 209. 1865; in DC. Prodr. 15(2): 1092. 1866 (non *J. ciliata*
Sessé, 1794).
J. longipedunculata Pax & Hoffm. Pflanzenreich 147, Addit. VII (Heft 85): 191. 1924 (non *J.*
longepedunculata Brandegee, 1920 [= *Cnidoscolus* sp.]).—*J. ciliata* var. *longipedunculata*
(Pax & Hoffm.) Macbride, Publ. Field Mus. Nat. Hist., Bot. Ser. 13(3A): 159. 1951.
J. hoffmanniae Croiz. J. Arnold Arb. 24: 168. 1943.
Type locality: Huánuco, Peru.

bartlettii Wilbur, J. Elisha Mitchell Sci. Soc. 70: 99. 1954. (II. 1.)
Type locality: Autlán, Jalisco, Mexico.

batawe Pax: see **stuhlmannii**

baumii Pax in Warburg, Kunene-Zambesi Exped. 283. 1903; Pflanzenreich 147 (Heft 42): 64. 1910; Hutchinson, Fl. Trop. Afr. 6(1): 782. 1913. (I. 5. A.)
Type locality: between Lionspan and the Kunene River, Angola.

berlandieri Torr.: see **cathartica**

berteri Spreng. Syst. Veg. 3:76.1826; Pax, Pflanzenreich 147 (Heft 42): 111. 1910 [nom. dubium].

bornmuelleri Pax, Pflanzenreich 147 (Heft 42): 133. 1910; McVaugh, Bull.Torrey Bot. Club 72: 280. 1945;
Allem, Rev. Bras. Biol. 37: 209. 1977. (I. 1. A.)
Type locality: Estancia Luiz, Rio Grande do Sul, Brazil.

brachyadenia Pax & Hoffm.: see **zeyheri**

brachypoda Pax, Pflanzenreich 147 (Heft 42): 74. 1910;[2] = *J. isabellii* X *dissecta* ? (see Lourteig & O'Donnell, Lilloa 9: 140. 1943; Allem, Rev. Bras. Biol. 37: 209. 1977).

breviloba (Morong) Pax & Hoffm. Pflanzenreich 147, Addit. VI (Heft 68): 38. 1919; McVaugh, Bull. Torrey Bot. Club 72: 280. 1945. (I. 1. A.)
Syn.: *J. gossypifolia* var. *breviloba* Morong, Ann. N.Y. Acad. Sci. 7: 219. 1892.—*J. ribifolia* var. *brevi-*
loba (Morong) Pax, Pflanzenreich 147 (Heft 42): 28. 1910; Lourt. & O'Donn. Lilloa 9: 38. 1943.
Type locality: Asunción, Paraguay.

---

2. Although he gave it a binomial, it seems that Pax did not intend to propose *J. brachyloba* as a distinct species but rather to indicate it as a hybrid taxon; under present nomenclature it could perhaps be
referred to as *J. X brachyloba*. The same holds true for *J. induta* and *J. transiens*.

brockmanii Hutchinson, Kew Bull. 1911: 360. 1911; Fl. Trop. Afr. 6(1): 787. 1913. (I. 1. B.)
  Type locality: Golis Range, Somali Republic.
brockmanii var. *leiosepala* Chiov.: see **trifida**

campestris S. Moore, J. Linn. Soc. Bot. 40: 196. 1911; Hutchinson, Fl. Trop. Afr. 6(1): 785. 1913. (I. 5. A?)
  Type locality: Umswirizwi Flats, Mozambique.
canescens (Benth.) Muell. Arg. in DC. Prodr. 15(2): 1079. 1866; Pax, Pflanzenreich 147 (Heft 42): 84, fig. 32. 1910; Dehgan & Webster, Madroño 25:33. 1978. (II. 3. B.)
  Syn.: *Mozinna canescens* Benth. Bot. Voy. Sulphur 52, t. 25. 1844.
  Type locality: Magdalena Bay, Baja California Sur, Mexico.
capensis (L.f.) Sond. Linnaea 23: 118. 1850; Baill. Adansonia 3: 149. 1862; Muell. Arg. in DC. Prodr. 15 (2): 1084. 1866; Sim, For. Fl. Cape Colony 310, t. 16, fig. 9. 1907; Pax, Pflanzenreich 147 (Heft 42): 54. 1910; Prain, Fl. Cap. 5(2): 421. 1920. (I. 5. B.)
  Syn.: *Croton capensis* L. f. Suppl. Pl. 422. 1781.
  Type locality: Cape of Good Hope, South Africa.
cardiophylla (Torr.) Muell. Arg. in DC. Prodr. 15(2): 1079. 1866; Pax, Pflanzenreich 147 (Heft 42): 85. 1910; Standley, Contr. U.S. Nat. Herb. 23: 639. 1923; McVaugh, Bull. Torrey Bot. Club 72: 288. 1945; Shreve & Wiggins, Veg. Fl. Sonoran Desert 1: 802. 1964. (II. 4.)
  Syn.: *Mozinna cardiophylla* Torr. Bot. U.S. Mex. Bound. 198. 1858.
  Type locality: "near Tucson and Sierra Verde, Sonora."
cathartica Terán & Berl. Mem. Comision Limites 9. 1832; I. M. Johnston, Contr. Gray Herb. 70: 89. 1924; McVaugh, Torrey Bot. Club 72: 277. 1945; Correll & Johnston, Fl. Texas 953. 1970. (I. 4.)
  Syn.: *J. berlandieri* Torr. Bot. U. S. Mex. Bound. 198. 1858; Muell. Arg. in DC. Prodr. 15(2): 1090. 1866; Pax, Pflanzenreich 147 (Heft 42): 41, fig. 14. 1910.—*Adenoropium berlandieri* (Torr.) Small, Addisonia 11: 55. 1927.
  Type locality: Tamaulipas, Mexico [Texas].
catingae Ule, Bot. Jahrb. Syst. 42: 218. 1908; Pax, Pflanzenreich 147 (Heft 42): 31, fig. 9. 1910. (I. 1?)[3]
  Type locality: Tambury, Bahia, Brazil.
cervicornis Susseng. Trans. Rhodesian Sci. Assoc. 43: 95. 1954. (I. 5. A?)
  Type locality: Rhodesia.
chevalieri Beille, Bull. Soc. Bot. France 55, Mém. 8: 83. 1908; Pax, Pflanzenreich 147 (Heft 42): 36. 1910; Hutchinson, Fl. Trop. Afr. 6(1): 788. 1913; Hutch. & Dalz. Fl. W. Trop. Afr. ed. 2, 1: 397. 1958. (I. 1. B.)
  Syn.: *J. lobata* γ *senegalensis* Muell. Arg. in DC. Prodr. 15(2): 1086. 1866.—*J. glauca* var. *senegalensis* (Muell. Arg.) Hiern, Cat. Afr. Pl. Welw. 4: 969. 1900.—*J. lobata* ssp. *senegalensis* (Muell. Arg.) Pax, Pflanzenreich 147 (Heft 42): 33. 1910.
  Type locality: Sénégal [Mauritania].
ciliata Sessé in Cerv. Gaz. Lit. Mex. 3: supl. 4. 1794; McVaugh, Bull. Torrey Bot. Club 72: 35. 1945. (II. 2.)
  Syn.: *J. olivacea* Muell. Arg. Linnaea 34: 207. 1865; in DC. Prodr. 15(2): 1078. 1866; Pax, Pflanzenreich 147 (Heft 42): 76. 1910.—*J. grandifrons* I. M. Johnston, Contr. Gray Herb. 68: 89. 1923.
  Type locality: Mexico [Puebla to Oaxaca].
ciliata Muell. Arg.: see **augustii**
cinerea (Ortega) Muell. Arg. in DC. Prodr. 15(2): 1078. 1866; Pax, Pflanzenreich 147 (Heft 42): 85. 1910; Standley, Contr. U. S. Nat. Herb. 23: 638. 1923; McVaugh, Bull. Torrey Bot. Club 72: 288. 1945; Shreve & Wiggins, Veg. Fl. Sonoran Desert 1: 804. 1964; Dehgan & Webster, Madroño, 25:32. 1978. (II. 3. B.)
  Syn.: *Mozinna cinerea* Ortega, Nov. Rar. Pl. Hort. Matr. 8: 107. 1799.
  Type locality: Mexico [Baja Cal., Sonora to Sinaloa].
clavuligera Muell. Arg. Linnaea 34: 209. 1865; in DC. Prodr. 15(2): 1086. 1866; Pax, Pflanzenreich 147 (Heft 42): 28. 1910; McVaugh, Bull. Torrey Bot. Club 72: 280. 1945; Macbride, Field Mus. Nat. Hist., Bot. Ser. 13(3A): 160. 1951. (I. 1. A.)
  Type locality: Sorata, Bolivia [Peru].

---

3. *Jatropha catingae* and *J. palmatifolia* differ from related taxa in having entire leaf lobes and androecia with only 6 stamens; they will probably be assigned to a separate subsection when they become better known.

cluytioides Pax & Hoffm.: see **lagarinthoides** var. **cluytioides**

coccinea Link: see **integerrima** var. **integerrima**

confusa Hutchinson, Kew Bull. 1911: 361. 1911; Fl. Trop. Afr. 6(1): 791. 1913. (I. 1. B.)

 Syn.: *J. glauca* var. *senegalensis* sensu Hiern, Cat. Afr. Pl. Welw. 4: 969. 1900 (non Muell. Arg. 1866).

 Type locality: Mossamedes, Angola.

cordata (Ortega) Muell. Arg. in DC. Prodr. 15(2): 1078. 1866; Pax, Pflanzenreich 147 (Heft 42): 85. 1910;
 Standley, Contr. U. S. Nat. Herb. 23: 638. 1923; McVaugh, Bull. Torrey Bot. Club 72: 290. 1945;
 Shreve & Wiggins, Veg. Fl. Sonoran Desert 1: 803. 1964. (II. 3. A.)

 Syn.: *Mozinna cordata* Ortega, Nov. Rar. Pl. Hort. Matr. 8: 107. 1799.

  Loureira glandulosa Cav. Icon. Descr. Pl. 5: 18, t. 430. 1799.

 Type locality: Mexico [Son. and Chih. south to Jalisco].

costaricensis Webster & Poveda, Brittonia, 30:265. 1978. (II. 2.)

 Type locality: Playa del Coco, Guanacaste, Costa Rica.

crinita Muell. Arg. Linnaea 34: 107. 1865; in DC. Prodr. 15(2): 1079. 1866; Pax, Pflanzenreich 147 (Heft
 42): 58. 1910; Hutchinson, Fl. Trop. Afr. 6(1): 781. 1913. (I. 3.)

 Type locality: Zanzibar, Tanzania.

cuneata Wiggins & Rollins, Contr. Dudley Herb. 3: 272. 1943; McVaugh, Bull. Torrey Bot. Club 72: 287.
 1945; Shreve & Wiggins, Veg. Fl. Sonoran Desert 1: 802. 1964. (II. 4.)

 Type locality: Sonora, Mexico [Arizona, Baja California].

cuneifolia Sessé & Moc. Fl. Mex. ed. 2, 224. 1894: see **dioica** var. **dioica.**

curcas L. Sp. Pl. 1006. 1753; Jacq. Hort. Vindob. 3: 36, t. 63. 1776; Griseb. Fl. Br. W. Ind. 36. 1859;
 Muell. Arg. in DC. Prodr. 15(2): 1080. 1866; in Mart. Fl. Bras. 11(2): 487, t. 68. 1874; Hook. f. Fl.
 Brit. Ind. 5: 383. 1887; Pax in Engler & Prantl. Natürl. Pflanzenfam. ed. 1, 3(5): 75. 1890; in Engler,
 Pflanzenw. Ost-Afr. C: 240. 1895; Cook & Collins, Contr. U. S. Nat. Herb. 8: 171, t. 42. 1903; Urban,
 Symb. Ant. 4: 349. 1905; Pax, Pflanzenreich 147 (Heft 42): 77, fig. 30. 1910; Hutchinson, Fl. Trop.
 Afr; 6(1) 791. 1913; Fawc. & Rendle, Fl. Jam. 4(2): 310 fig. 103. 1920; Prain, Fl. Cap. 5(2): 420.
 1920; Standley, Contr. U. S. Nat. Herb. 23: 640. 1923; Lourt. & O'Donn., Lilloa 9: 119. 1943;
 McVaugh, Bull. Torrey Bot. Club 72: 283. 1945; Standley & Steyermark, Fieldiana Bot. 24(6): 127.
 1949; Macbride, Field Mus. Nat. Hist., Bot. Ser. 13(3A): 160. 1951; Alain, Fl. Cuba 3: 76. 1951; Wil-
 bur, J. Elisha Mitchell Sci. Soc. 70: 94. 1954; Schnell, Icon. Pl. Afr. 5: 110. 1960; Webster & Burch,
 Ann. Missouri Bot. Gard. 54: 23. 1968; Airy Shaw, Kew Bull. 26: 283. 1971; Little et al. Trees Puerto
 Rico & Virgin I. 2 [USDA Handb. 449]: 416, t. 444. 1974. (II. 1.)

 Syn.: *Curcas purgans* Medik. Ind. Pl. Hort. Manhem. 1: 90. 1771; Baill. Étud. Gen. Euphorb. 314.
  1858.

  Ricinus americanus Miller, Gard. Dict. ed. 8. 1768.

  Castiglionia lobata Ruiz & Pavon, Fl. Peruv. Prodr. 139, t. 37. 1794.

  Jatropha edulis Cerv. Gaz. Lit. Mex. 3: supl. 4. 1794.

  J. acerifolia Salisb. Prodr. Chapel Allerton 389. 1796.

  Ricinus jarak Thunb. Fl. Javan. 23. 1825.

  Curcas adansoni Endl. ex Heynh. Nomencl. 176. 1840.

  Curcas indica A. Rich in Sagra, Hist. Fis. Pol. Nat. Cuba 3: 208. 1853.

  ? *Jatropha yucatanensis* Briq. Ann. Cons. Jard. Genève 4: 230. 1900; Standley, Contr. U. S. Nat.
   Herb. 23: 640. 1923; McVaugh, Bull. Torrey Bot. Club 72: 35. 1945.

  Curcas curcas (L.) Britton & Millsp. Bahama Fl. 225. 1920.

 Type locality: "America calidiore" [native to America, now circumtropical].

curcas var. *rufus* McVaugh: see **mcvaughii**

decumbens Pax & Hoffm. Pflanzenreich 147, Addit. V (Heft 63): 398. 1914. (I. 5. A?)

 Type locality: southwest Africa.

dioica Sessé in Cerv. Gaz. Lit. Mex. 3: supl. 4. 1794; McVaugh, Bull. Torrey Bot. Club 72: 36. 1945; Johnston & Warnock, Southw. Nat. 8: 122. 1963. (II. 4.)

 var. **dioica.**

 Syn.: *Loureira cuneifolia* Cav. Icon. Descr. Pl. 5: 17, t. 429. 1799.—*Curcas cuneifolium* (Cav.) Baillon,
  Étud. Gen. Euphorb. 315. 1858.—*Jatropha cuneifolia* Sessé & Moc. Fl. Mex. ed. 2, 224. 1894.

 Syn.: *Loureira cuneifolia* Cav. Icon. Descr. Pl. 5: 17, t. 429. 1799.—*Curcas cuneifolium* (Cav.) Baillon,
   Etud. Gen. Euphorb. 315. 1858.—*Jatropha cuneifolia* Sessé & Moc. Fl. Mex. ed. 2, 224. 1894.

Mozinna spathulata Ortega, Nov. Rar. Pl. Hort. Matr. 8: 105, t. 13. 1799.–*Jatropha spathulata* (Ortega) Muell. Arg. in DC. Prodr. 15(2): 1081. 1866; Pax, Pflanzenreich 147 (Heft 42): 81, fig. 31. 1910; Standley, Contr. U. S. Nat. Herb. 23: 637. 1923.

    Mozinna spathulata var. *sessiliflora* Hook. Icon. Pl. 4: t. 357. 1841.–*Jatropha spathulata* β *sessiliflora* (Hook.) Muell. Arg. in DC. Prodr. 15(2): 1082. 1866.–*Mozinna sessiliflora* (Hook.) Small, Fl. Southeastern U.S. 706. 1903.–*Jatropha dioica* var. *sessiliflora* (Hook.) McVaugh, Bull. Torrey Bot. Club 72: 37. 1945.

    Zimapania schiedeana Engler & Pax, Natürl. Pflanzenfam. ed. 1, 3(5): 119. 1891.

Type locality: Mexico [Oax. to Durango, Tam., N. L.; Texas].

    var. **graminea** McVaugh, Bull. Torr. Bot. Club 72: 39. 1945; Johnston & Warnock, Southw. Nat. 8: 123. 1963.

    Type locality: Jimulco, Coahuila, Mexico [Zacatecas to Chihuahua; Texas].

dissecta (Chod. & Hassl.) Pax, Pflanzenreich 147 (Heft 42): 72, fig. 28. 1910; McVaugh, Bull. Torrey Bot. Club 72: 280. 1945; Allem, Rev. Bras. Biol. 37: 209. 1977. (I. 1. A.)

    Syn.: *J. gossypifolia* ssp. *heterophylla* var. *dissecta* Chod. & Hassler, Bull. Herb. Boissier II. 5: 611. 1905.

    Type locality: Paraguay.

divaricata Swartz, Nov. Gen. Spec. Pl. Prodr. 98. 1788. Fl. Ind. Occ. 2: 1158. 1800; Muell. Arg. in DC. Prodr. 15(2): 1077. 1866; Pax, Pflanzenreich 147 (Heft 42): 74. 1910; Fawc. & Rendle, Fl. Jam. 4(2): 314. 1920; McVaugh, Bull. Torrey Bot. Club 72: 286. 1945; Adams, Fl. Pl. Jam. 416. 1972. (I. 6. B.)

    Syn.: *Adenoropium divaricatum* (Swartz) Pohl, Pl. Bras. Icon. Descr. 1: 14. 1827.

    Type locality: Jamaica.

divergens Baillon: see **mollissima** var. **divergens**

diversifolia A. Rich.: see **integerrima**

edulis Cerv.: see **curcas**

eglandulosa Pax, Pflanzenreich 147 (Heft 42): 63. 1910; McVaugh, Bull. Torrey Bot. Club 72: 280. 1945; Allem, Rev. Bras. Biol. 37: 220. 1977. (I. 1. A.)

    Syn.: *J. elliptica* var. *guaranitica* Chodat & Hassler, Bull. Herb. Boissier II. 5: 611. 1905.

    Type locality: Apa River, Paraguay.

elegans (Pohl) Kl.: see **gossypiifolia** var. **elegans**

ellenbeckii Pax, Bot. Jahrb. Syst. 33: 284. 1902; Pflanzenreich 147 (Heft 42): 58, fig. 22. 1910; Hutchinson, Fl. Trop. Afr. 6(1): 797. 1913. (I. 3.)

    Type locality: Harar, Ethiopia.

elliptica (Pohl) Muell. Arg. in Mart. Fl. Brasil. 11(2): 489. 1874; Pax, Pflanzenreich 147 (Heft 42): 62. 1910; McVaugh, Bull. Torrey Bot. Club 72: 280. 1945; Allem, Rev. Bras. Biol. 37: 220. 1977. (I. 1. A.)

    Syn.: *Adenoropium ellipticum* Pohl, Pl. Bras. Icor. Descr. 1: 13, t. 9. 1827.

    Jatropha opifera Mart. Reise Brasil 2: 548. 1828.

    J. lacerti Silva Manso, Enum. Subst. Brazil 8. 1836.

    J. officinalis Mart. ex Baillon, Adansonia I. 4: 266. 1864; Muell. Arg. in DC. Prodr. 15(2): 1089. 1866.

Type locality: Goyaz, Minas Gerais, Brazil.

elliptica var. *guaranitica* Chodat & Hassler: see **eglandulosa**

erythropoda Pax & Hoffm. Pflanzenreich 147 (Heft 42): 66. 1910; Prain, Fl. Cap. 5(2): 422. 1920. (I. 5. A.)

    var. **erythropoda**

        Type locality: Neitsas, Omaheke, Southwest Africa.

    var. **hirtula** Pax & Hoffm. Pflanzenreich 147, Addit. V (Heft 63): 399. 1914.

        Type locality: Epata, Omaheke, Southwest Africa.

excisa Griseb. Abh. Konigl. Ges. Wiss. Göttingen 19: 94. 1874; Pax, Pflanzenreich 147 (Heft 42): 30. 1910; Lourt. & O'Donn. Lilloa 9: 121, t. 8. 1943; McVaugh, Bull. Torrey Bot. Club 72: 280. 1945. (I. 1. A.)

    var. **excisa**

        Type locality: Catamarca, Argentina.

    var. **pubescens** Lourt. & O'Donn. Lilloa 9: 123, fig. 11, t. 9. 1943.

        Type locality: Oran, Salta, Argentina.

    var. **viridiflora** Lourt. & O'Donn. Lilloa 9: 125, fig. 12, t. 10. 1943.

        Type locality: Rosario de la Frontera, Salta, Argentina.

ferox Pax, Annuario Reale Ist. Bot. Roma 6: 185. 1896; Pflanzenreich 147 (Heft 42): 56. 1910; Hutchinson, Fl. Trop. Afr. 6(1): 780. 1913; Chiov. Fl. Somala 1: 306. 1929. (I. 3.)
Type locality: Merehan, Somali Republic.

fissispina Pax, Bot. Jahrb. Syst. 43: 83. 1909; Pflanzenreich 147 (Heft 42): 58, fig. 23. 1910; Hutchinson, Fl. Trop. Afr. 6(1): 798. 1913; Agnew, Upland Kenya Wild Flowers 218. 1974. (I. 3.)
Type locality: Ol Dongo, Masai Steppe, Tanzania [Kenya].

flabellifolia Pax & Hoffm.: see **paxii**

flavovirens Pax & Hoffm. Pflanzenreich 147 (Heft 42): 30. 1910; Lourt. & O'Donn. Lilloa 9:125. 1943; McVaugh, Bull. Torrey Bot. Club 72: 280. 1945. (I. 1. A; = **J. excisa?**)
Type locality: Loma Clavel, Gran Chaco, Paraguay.

fremontioides Standley, Field Mus. Nat. Hist., Bot. Ser. 22: 37. 1940; McVaugh, Bull. Torrey Bot. Club 72: 286. 1945. (II. 3. C.)
Type locality: Ixtepec, Oaxaca, Mexico.

gallabatensis Schweinf. Verh. Zool.-Bot. Ges. Wein 18: 661. 1868; Pax, Pflanzenreich 147 (Heft 42): 69. 1910; Hutchinson, Fl. Trop. Afr. 6(1): 794. 1913. (I. 5. A.)
Type locality: Gallabat, Sudan.

gaumeri Greenman, Publ. Field Columbian Mus., Bot. Ser. 2: 256. 1907; Pax, Pflanzenreich 147 (Heft 42): 133. 1910; McVaugh, Bull. Torrey Bot. Club 72: 286. 1945; Standley & Steyermark, Fieldiana Bot. 24(6): 128. 1949. (I. 6. B?)
Type locality: Izamal, Yucatán, Mexico [Belize; Guatemala].

giffordiana Dehgan & Webster, Madroño 25:30. 1978. (II. 3. B.)
Type locality: Cabo San Lucas, Baja California Sur, Mexico.

glabrescens Pax & Hoffm.: see **hirsuta** var. **glabrescens**

glandulifera Roxb. Fl. Ind. 3: 688. 1832; Muell. Arg. in DC. Prodr. 15(2): 1084. 1866; Hook. f. Fl. Brit. Ind. 5: 382. 1887; in Trimen, Handb. Fl. Ceylon 4: 45. 1898; Cooke, Fl. Pres. Bombay 93. 1906; Pax, Pflanzenreich 147 (Heft 42): 31. 1910. (I. 1. B.)
Syn.: *Adenoropium roxburghii* Kostel. Med. Pharm. Fl. 5: 1750. 1836.
Type locality: vicinity of Calcutta, India.

glandulosa Vahl: see **pelargoniifolia**

glauca Vahl, Symb. Bot. 1: 78. 1790; Schweinf. Beitr. Fl. Aethiop. 37. 1867; Andrews, Fl. Pl. Anglo-Egypt. Sudan 2: 83. 1952. (I. 5. B.)
Syn.: *Croton lobatus* sensu Forsk. Fl. Aegypt. Arab. 162. 1775 (non L. 1753).—*Jatropha lobata* Muell. Arg. in DC. Prodr. 15(2): 1085. 1866; Pax, Pflanzenreich 147 (Heft 42): 32. 1910; Hutchinson, Fl. Trop. Afr. 6(1): 788. 1913.
Adenoropium glaucum (Vahl) Pohl, Pl. Bras. Icon. Descr. 1: 15. 1827.—*Jatropha lobata* ssp. *glauca* (Vahl) Pax, Pflanzenreich 147 (Heft 42): 32, fig. 10. 1910.—*J. lobata* var. *glauca* (Vahl) Pax ex Blatter, Rec. Bot. Surv. India 8: 435. 1923.
Jatropha lobata β *richardiana* Muell. Arg. in DC Prodr. 15(2): 1086. 1866.
Type locality: Mor, Arabia [northeast Africa].

glauca Griseb.: see **angustifolia** var. **glauca**

glaucovirens Pax & Hoffm.: see **integerrima**

gossypiifolia L. Sp. Pl. 1:1006. 1753; Jacq. Icon. Pl. Rar. 3: t. 623. 1788; Swartz, Observ. Bot. 366. 1791; Lodd. Bot. Cab. 2: t. 117. 1818; Griseb. Fl. Br. W. Ind. 36. 1859; Muell. Arg. in DC. Prodr. 15(2): 1086. 1866; in Mart. Fl. Brasil. 11(2): 491. 1874; Hook. f. Fl. Br. Ind. 5: 383. 1887; Pax in Engler, Pflanzenw. Ost-Afr. C: 240. 1895; Urban, Symb. Ant. 4:350. 1905; Pax, Pflanzenreich 147 (Heft 42): 26. 1910; Hutchinson, Fl. Trop. Afr. 6(1): 783. 1913; Holm, Merck's Rept. 24: 165. 1915; Fawc. & Rendle, Fl. Jam. 4(2): 312. 1920; Standley, Contr. U.S. Nat. Herb. 23: 637. 1923; Hutchinson & Dalz. Fl. W. Trop. Afr. ed. 1, 297. 1928; Dalziel, Useful Pl. W. Trop. Afr. 148. 1937; McVaugh, Bull. Torrey Bot. Club 72: 281. 1945; Standley & Steyermark, Fieldiana Bot. 24(6): 128. 1949; Macbride, Field Mus. Nat. Hist., Bot. Ser. 13(3A): 161. 1951; Alain, Fl. Cuba 3: 75. 1953; Schnell, Icon. Pl. Afr. 5: 111. 1960; Irvine, Woody Pls. Ghana 236. 1961; Gooding et al. Fl. Barbados 255. 1965; Webster & Burch, Ann. Missouri Bot. Gard. 54: 235, fig. 5. 1968; Airy Shaw, Kew Bull. 26: 283. 1971. (I. 1. A.)
    var. **gossypiifolia**
        Syn.: *J. staphysagrifolia* Miller, Gard. Dict. ed. 8. 1768.—*J. gossypiifolia* α *staphysagriaefolia* Muell. Arg. in DC. Prodr. 15(2): 1087. 1866.

Adenoropium gossypifolium (L.) Pohl, Pl. Bras. Icon. Descr. 1: 16. 1827.
Adenoropium jacquini Pohl, op. cit. 15.—*Jatropha jacquini* (Pohl) Baillon, Adansonia I.
4: 268. 1864.
Type locality: American Tropics [circumtropical].
var. **elegans** (Pohl) Muell. Arg. in DC. Prodr. 15(2): 1087. 1866; in Mart. Fl. Brasil. 11(2): 492. 1874;
Chodat & Hassler, Bull. Herb. Boissier II. 5: 611. 1905; Pax, Pflanzenreich 147 (Heft 42): 26.
1910; Backer & Bakh. f. Fl. Java 1: 494. 1963.
Syn.: *Adenoropium elegans* Pohl, Pl. Bras. Icon. Descr. 1: 15. 1827.—*Jatropha elegans* (Pohl) Kl.
in Seem. Bot. Voy. Herald 102. 1853.
Type locality: Ilhéus, Bahia, Brazil [tropical America and Africa]
gossypiifolia var. *breviloba* Morong: see **breviloba**
gossypiifolia ssp. *heterophylla* Chodat & Hassler: see **isabellii**
gossypiifolia ssp. *heterophylla* var. *dissecta* Chodat & Hassler: see **dissecta**
gossypiifolia ssp. *heterophylla* var. *dissecta* f. *induta* Chodat & Hassler: see **induta**
gossypiifolia ssp. *heterophylla* var. *grandifolia* Chodat & Hassler: see **isabellii** var. **grandifolia**
gossypiifolia ssp. *heterophylla* var. *guaranitica* Chodat & Hassler: see **isabellii** var. **guaranitica**
gossypiifolia ssp. *heterophylla* var. *palmata* Chodat & Hassler: see **isabelli** var. **palmata**
gossypiifolia ssp. *heterophylla* var. *rhombifolia* Chodat & Hassler: see **isabellii** var. **rhombifolia**
gossypiifolia var. *intermedia* Chodat & Hassler: see **intermedia**
gossypiifolia var. *isabellii* (Muell. Arg.) Chodat & Hassler f. *glabrata*, f. *latifolia* Chodat & Hassler: see **intermedia**
grandifrons I. M. Johnston: see **ciliata**
grossidentata Pax & Hoffm. Pflanzenreich 147, Addit. V(Heft 63): 398. 1914; Lourt. & O'Donn. Lilloa 9:
127, fig. 13, t. 11. 1943; McVaugh, Bull. Torrey Bot. Club 72: 277. 1945; Allem, Rev. Bras. Biol. 37:
220. 1977. (I. 4.)
Syntype localities: Nueva Pompey, between Los Labalos and Tres Tobas, Paraguay; Ipawassu, Bolivia
[Argentina].
guaranitica Speg. Anales Soc. Ci. Argent. 16: 93. 1883; Pax, Pflanzenreich 147 (Heft 42): 29. 1910; Mc-
Vaugh, Bull. Torrey Bot. Club 72: 280. 1945. (I. 1. A.)
Type locality: Tucurupucu, Parana River, Paraguay.
harms
harmsiana Mattf.: see **neopauciflora**
hastata Jacq.: see **integerrima** var. **hastata**
hernandiifolia Vent. Jard. Malm. 1: 52. 1803; Muell. Arg. in DC. Prodr. 15(2): 1077. 1866; Urban, Symb.
Ant. 4: 349. 1905; Pax, Pflanzenreich 147 (Heft 42): 74. 1910; Fawc. & Rendle, Fl. Jam. 4(2): 314.
1920; McVaugh, Bull. Torrey Bot. Club 72: 286. 1945; Adams, Fl. Pl. Jam. 416. 1972; Little et al.
Trees Puerto Rico & Virgin I. 2 [USDA Handb. 449]: 418, t. 445. 1974. (I. 6. B.)
var. **hernandiifolia**
Syn.: *Adenoropium hernandiaefolium* (Vent.) Pohl, Pl. Brasil. Icon. Descr. 1: 14. 1827.—*Curcas portoricensis* Baillon, Étud. Gen. Euphorb. 314. 1858. — *Curcas hernandifolius* (Vent.)
Britton, Sci. Surv. Puerto Rico Virgin I. 5: 484. 1924.
Loureira peltata Desf. Tabl. École Bot., Paris ed. 3, 411. 1829.—*Mozinna peltata* (Desf.)
Steud. Nomencl. Bot. ed. 2, 1: 483. 1840. — *Curcas peltata* (Desf.) Baill. Étud. Gen. Euphorb. 315. 1858.—*Jatropha hernandiaefolia* var. *peltata* (Desf.) Pax, Pflanzenreich 147
(Heft 42): 75. 1910.
Jatropha heterophylla Sessé & Moc. Fl. Mex. ed. 2, 224. 1894.
Type locality: Puerto Rico [Haiti; Jamaica].
var. **portoricensis** (Millsp.) Urban, Symb. Ant. 4: 349. 1905.
Syn.: *Jatropha portoricensis* Millsp. Publ. Field Columbian Mus., Bot. Ser. 2: 59, t. 62. 1900.
Type locality: Puerto Rico.
var. **epeltata** Pax, Pflanzenreich 147 (Heft 42): 76. 1910.
Type locality: Haiti.
heterophylla Heyne: see **heynei**
heterophylla Sessé & Moc.: see **hernandiifolia**
heterophylla Pax: see **variifolia**
heynei Balakr. Bull. Bot. Surv. India 3: 40. 1962. (I. 5. A?)

Syn.: *J. heterophylla* Heyne ex Hook. f. Fl. Brit. Ind. 5: 382. 1887 (non Steud. 1840); Pax, Pflanzenreich 147 (Heft 42): 70. 1910.

Type locality: Deccan, India.

hieronymii O. Ktze. Rev. Gen. Pl. 3(2): 287. 1898; Pax, Pflanzenreich 147 (Heft 42): 36. 1910; Lourt. &
O'Donn. Lilloa 9: 128, fig. 14, t. 12. 1943; McVaugh, Bull. Torrey Bot. Club 72: 277. 1945; Allem,
Rev. Bras. Biol. 37: 217. 1977. (I. 4.)

Type locality: Oran, Jujuy, Argentina [Bolivia].

hildebrandtii Pax, Bot. Jahrb. Syst. 19: 108. 1894; in Engler, Pflanzenw. Ost-Afr. C: 240. 1895; Pflanzenreich 147 (Heft 42): 35. 1910; Hutchinson, Fl. Trop. Afr. 6(1): 792. 1913; Pax in Engler & Drude, Veg.
Erde 9. III(2): 112. 1921. (I. 1. B.)

var. **hildebrandtii**

Type locality: Lamu Island, Kenya [Zanzibar, Tanzania, Comoro Islands].

var. **torrentis-lugardi** Radcliffe-Smith, Kew Bull. 28: 283. 1973.

Type locality: Yatta Plateau, Kenya.

hintonii Wilbur, J. Elisha Mitchell Sci. Soc. 70: 95. 1954. (II. 1.)

Type locality: Zacatecas, Mexico.

hippocastanifolia Croizat, J. Arnold Arb. 24: 168. 1943; McVaugh, Bull. Torrey Bot. Club 72: 281. 1945.
(I. 1. A.)

Type locality: Oruro, Bolivia.

hirsuta Hochst. Flora 28: 82. 1845; Muell. Arg. in DC. Prodr. 15(2): 1088. 1866; Pax, Pflanzenreich 147
(Heft 42): 62, fig. 24. 1910; Prain, Fl. Cap. 5(2): 434. 1920. (I. 5. A.)

var. **hirsuta**

Type locality: Umlass River, Natal, South Africa.

var. **glabrescens** (Pax & Hoffm.) Prain, Fl. Cap. 5(2): 424. 1920.

Syn.: *J. glabrescens* Pax & Hoffm. Pflanzenreich 147 (Heft 42): 62. 1910.

Type locality: Clairmont, Natal, South Africa.

var. **oblongifolia** Prain, Fl. Cap. 5(2): 424. 1920.

Syntype localities: Transvaal, Swaziland, South Africa.

humboldtiana McVaugh, Bull. Torrey Bot. Club 72: 35. 277. 1945; Macbride, Field Mus. Nat. Hist., Bot.
Ser. 13(3A): 161. 1951. (I. 4.)

Syn.: *J. peltata* H.B.K. Nov. Gen. Sp. 2: 104. 1817 (non Sessé 1794 nec Wright 1870); Muell. Arg. in
DC. Prodr. 15(2): 1092. 1866; in Mart. Fl. Brasil. 11(2): 491. 1874; Pax, Pflanzenreich 147 (Heft
42): 43. 1910.–*Adenoropium peltatum* (H.B.K.) Pohl, Pl. Bras. Icon. Descr. 1: 16. 1827.

Type locality: Jaen de Bracamoros, Cajamarca, Peru.

humilis N. E. Br.: see **seineri**

induta (Chodat & Hassler) Pax, Pflanzenreich 147 (Heft 42): 72. 1910: = *J. isabellii* X *dissecta?* (see Lourteig & O'Donnell, Lilloa 9: 140. 1943; Allem, Rev. Bras. Biol. 37: 209. 1977).

integerrima Jacq. Stirp. Sel. Amer. 265, t. 183 f. 47. 1763; Sims, Bot. Mag. 36: t. 1464. 1812; Pax, Pflanzenreich 147 (Heft 42): 50, fig. 19. 1910; McVaugh, Bull. Torrey Bot. Club 72: 274. 1945; Alain, Fl.
Cuba 3: 77. 1953; Webster & Burch, Ann. Missouri Bot. Gard. 54: 237. 1968; Airy Shaw, Kew Bull.
26: 284. 1971. (I. 6. A.)

var. **integerrima**[4]

Syn.: *J. acuminata* Desr. in Lam. Encycl. Meth. 4: 10. 1797; Vent. Jard. Malm. 1: 52. 1803.

J. pandurifolia Andr. Bot. Repos. 4: t. 267. 1799; Sims, Bot. Mag. 17: t. 604. 1803; Lodd.
Bot. Cab. 17: t. 1604. 1830; Muell. Arg. in DC. Prodr. 15(2): 1095. 1866; Pax, Pflanzenreich 147 (Heft 42): 49. 1910.–*Adenoropium pandurifolium* Pohl, Pl. Bras. Icon. Descr.
1: 14. 1827.–*Jatropha diversifolia* γ *panduraefolia* (Andr.) Gomez, Anal. Hist. Nat. Madrid 23: 51. 1894.

J. coccinea Link, Enum. Pl. Hort. Berol. 2: 406. 1822.–*J. pandurifolia* var. *coccinea* (Link)
Pax, Pflanzenreich 147 (Heft 42): 50. 1910.

---

4. It is uncertain whether all of the synonyms listed here refer to var. *integerrima;* some may belong
under var. *hastata;* but in any event, there is some doubt whether var. *hastata* can be maintained as a distinct variety.

J. pauciflora Griseb. Abh. Konigl. Ges. Wiss. Göttingen 12: 170. 1865; Muell. Arg. in DC.
   Prodr. 15(2): 1092. 1866; Pax, Pflanzenreich 147 (Heft 42): 51. 1910.—*J. diversifolia* δ
   pauciflora (Griseb.) Maza, Anal. Hist. Nat. Madrid 23: 51. 1894.
J. moluensis Sessé & Moc. Fl. Mex. ed. 2, 224. 1894; McVaugh, Bull. Torrey Bot. Club 72:
   33. 1945.
J. pandurifolia var. *latifolia* Pax, Pflanzenreich 147 (Heft 42): 50. 1910.
   Type locality: Cuba [Hispaniola].
var. **hastata** (Jacq.) Fosb. Rhodora 78: 102. 1976.
   Syn.: *J. hastata* Jacq. Stirp. Sel. Amer. 256, t. 173, f. 54. 1763; Griseb. Fl. Br. W. Ind. 36. 1859;
      Standley & Steyermark, Fieldiana Bot. 24(6): 129. 1949.—*Adenoropium hastatum*
      (Jacq.) Britton & Wilson, Sci. Surv. Puerto Rico & Virgin I. 5: 485. 1924.
   J. diversifolia A. Rich. in Sagra, Hist. Fis. Pol. Nat. Cuba 3: 207. 1853; Muell. Arg. in DC.
      Prodr. 15(2): 1094. 1866.
   Type locality: Cuba.
intercedens Pax, Pflanzenreich 147 (Heft 42): 31. 1910; McVaugh, Bull. Torrey Bot. Club 72: 280. 1945.
   (I. 1. A.)
   Type locality: near Bermejo, Bolivia.
intermedia (Chodat & Hassler) Pax, Pflanzenreich 147 (Heft 42): 63. 1910; McVaugh, Bull. Torrey Bot.
   Club 72: 280. 1945; Allem, Rev. Bras. Biol. 37: 209. 1977. (I. 1. A; = **J. isabellii?**)
   Syn.: *J. gossypifolia* var. *intermedia* Chodat & Hassler, Bull. Herb. Boiss. II. 5: 612. 1905.
      J. gossypifolia var. *isabellii* f. *glabrata*, f. *latifolia* Chodat & Hassler, loc. cit.
   Type locality: Itacurubi, Paraguay [Brazil].
isabellii Muell. Arg. in Mart. Fl. Bras. 11(2): 489. 1874; Pax, Pflanzenreich 147 (Heft 42): 71. 1910; Allem,
   Rev. Bras. Biol. 37: 209. 1977. (I. 1. A.)
   var. **isabellii**
      Type locality: Rio Grande do Sul, Brazil.
   var. **antisyphilitica** (Speg.) Pax, Pflanzenreich 147 (Heft 42): 72. 1910; Lourt. & O'Donn. Lilloa 9: 131,
      fig. 15. 1943; McVaugh, Bull. Torrey Bot. Club 72: 280. 1945.
      Syn.: *J. antisyphilitica* Speg. Anal. Soc. Ci. Argent. 16: 91. 1883.
      Type locality: Ituzaingo, Argentina [Paraguay].
   var. **cuneifolia** Pax, Pflanzenreich 147 (Heft 42): 71. 1910.
      Syn.: *J. gossypifolia* ssp. *heterophylla* var. *typica* Chodat & Hassler, Bull. Herb. Boiss. II. 5: 711.
         1905.
      Type locality: Cordillera de Altos, Paraguay.
   var. **grandifolia** (Chodat & Hassler) Pax, Pflanzenreich 147 (Heft 42): 71. 1910.
      Syn.: *J. gossypifolia* ssp. *heterophylla* var. *grandifolia* Chodat & Hassler, Bull. Herb. Boiss. II. 5: 612.
         1905.
      Type locality: Valenzuela, Paraguay.
   var. **guaranitica** (Chodat & Hassler) Pax, Pflanzenreich 147 (Heft 42): 71. 1910.
      Syn.: *J. gossypifolia* ssp. *heterophylla* var. *guaranitica* Chodat & Hassler, Bull. Herb. Boiss. II. 5: 612.
         1905.
      Type locality: Corrientes, Argentina.
   var. **palmata** (Chodat & Hassler) Pax, Pflanzenreich 147 (Heft 42): 71. 1910.
      Syn.: *J. gossypifolia* ssp. *heterophylla* var. *palmata* Chodat & Hassler Bull. Herb. Boiss. II. 5: 612.
         1905.
      Type locality: Valenzuela.
   var. **rhombifolia** (Chodat & Hassler) Pax, Pflanzenreich 147 (Heft 42): 72. 1910.
      Syn.: *J. gossypifolia* ssp. *heterophylla* var. *rhombifolia* Chodat & Hassler, Bull. Herb. Boiss. II. 5:
         612. 1905.
      Type locality: Capibary River, Paraguay.

jacquinii Baillon: see **gossypiifolia**

kamerunica Pax & Hoffm. Pflanzenreich 147, Addit. II (Heft 47): 102. 1911; Hutchinson, Fl. Trop. Afr.
   6(1): 795. 1913. (I. 1. B.)
   var. **kamerunica**
      Type locality: Dangadji, Cameroon.

var. **trochainii** Leandri, Bull. Soc. Bot. France 83: 525. 1936.
Type locality: Tambecounda, Senegal.
katharinae Pax, Pflanzenreich 147 (Heft 42): 28, fig. 8. 1910; McVaugh, Bull. Torrey Bot. Club 72: 280. 1945. (I. 1. A.)
Type locality: Caaguazu, Paraguay.
kilimanscharica Pax & Hoffm.: see **spicata**

lacerti Silva Manso: see **elliptica**
lagarinthoides Sond. Linnaea 23: 118. 1850; Muell. Arg. in DC. Prodr. 15(2): 1088. 1866; Pax, Pflanzen-reich 147 (Heft 42): 64, fig. 25. 1910; Prain, Fl. Cap. 5(2): 423. 1920. (I. 5. A.)
var. **lagarinthoides**
Type locality: Cape of Good Hope, South Africa [Transvaal].
var. **cluytioides** (Pax & Hoffm.) Prain, Fl. Cap. 5(2): 423. 1920.
Syn.: *J. cluytioides* Pax & Hoffm. Pflanzenreich 147 (Heft 42): 65. 1910.
J. latifolia var. *stenophylla* Pax, op. cit. 133.
Type locality: Transvaal, South Africa.
latifolia Pax, Bot. Jahrb. Syst. 23: 531. 1897; Pflanzenreich 147 (Heft 42): 61. 1910; Prain, Fl. Cap. 5(2): 424. 1920. (I. 5. A.)
var. **latifolia**
Type locality: Lydenburg, Transvaal, South Africa.
var. **angustata** Prain, Fl. Cap. 5(2): 424. 1920.
Type locality: Kalahari region, South Africa.
var. **swazica** Prain, loc. cit.
Type locality: Swaziland, South Africa.
latifolia var. *stenophylla* Pax: see **lagarinthoides** var. **cluytioides**
lobata (Forsk.) Muell. Arg.: see **glauca**
lobata var. *glauca* (Vahl) Blatter: see **glauca**
lobata var. *richardiana* Muell. Arg.: see **glauca**
lobata var. *senegalensis* Muell. Arg.: see **chevalieri**
lobata ssp. *aceroides* Pax & Hoffm.: see **aceroides**
longipedunculata Pax & Hoffm.: see **augustii**
luxurians (Pohl) Baillon: see **mollissima** var. **mollissima**

macrantha Muell. Arg. Linnaea 34: 209. 1865; in DC. Prodr. 15(2): 1082. 1866; Pax, Pflanzenreich 147 (Heft 42): 48, fig. 18. 1910; McVaugh, Bull. Torrey Bot. Club 72: 275. 1945; Macbride, Field Mus. Nat. Hist., Bot. Ser. 13(3A): 161. 1951. (I. 4.)
Type locality: Peru.
macrocarpa Griseb. Abh. Konigl. Ges. Wiss. Göttingen 19: 94. 1874; Pax, Pflanzenreich 147 (Heft 42): 47. 1910; Lourt. & O'Donn. Lilloa 9: 132, fig. 16, t. 13. 1943; McVaugh, Bull. Torrey Bot. Club 72: 277. 1945. (I. 4?)
Syn.: *J. multiflora* Pax & Hoffm. Pflanzenreich 147, Addit. VII (Heft 85): 1924.
Type locality: Catamarca, Argentina.
macrophylla Pax & Hoffm. Pflanzenreich 147 (Heft 42): 80. 1910; Hutchinson, Fl. Trop. Afr. 6(1): 792. 1913. (II. 1?)
Type locality: Malawi.
macrorhiza Benth. Pl. Hartweg. 8. 1839; Muell. Arg. in DC. Prodr. 15(2): 1087. 1866; Pax, Pflanzenreich 147 (Heft 42): 70. 1910; McVaugh, Bull. Torrey Bot. Club 72: 280. 1945; Johnston & Warnock, Southw. Nat. 8: 123. 1963; Shreve & Wiggins, Veg. Fl. Sonoran Desert 1: 801. 1964; Correll & Johnston, Fl. Texas 954. 1970. (I. 6. B.)
var. **macrorhiza**
Type locality: Guanajuato, Mexico [southwestern U.S. & northwestern Mexico].
var. **septemfida** Engelm. in Rothr. Rept. U.S. Geogr. Surv. W. 100th Merid. 6: 243. 1878.
Syn.: *J. arizonica* I. M. Johnston, Contr. Gray Herb. 67: 89. 1923.
Type locality: Arizona [Sonora, Chihuahua, Mexico].
mahafalensis Jumelle & Perrier, Bull. Econ. Madagascar 10: 180. 1910; Jumelle, Rev. Gen. Bot. 32: 121.

1920; Pax & Hoffm. Natürl. Pflazenfam. ed. 2, 19c: 160. 1931; Leandri, Cat. Pl. Madag. Euphorb. 48.
1935: probably not a *Jatropha.*[5]

maheshwarii Subram. & Nayar, Bull. Bot. Surv. India 6: 331, figs. 1-8. 1965: = *Dimorphocalyx beddomii*
(Benth.) Airy Shaw, fide Airy Shaw, Kew Bull. 27: 92. 1972.

malacophylla Standley, Proc. Biol. Soc. Wash. 37: 45. 1924; McVaugh, Bull. Torrey Bot. Club 72: 287.
1945. (II. 1.)
Syn.: *J. platanifolia* Standley, Publ. Field Mus. Nat. Hist., Bot. Ser. 22: 38. 1940; Shreve & Wiggins,
Veg. Fl. Sonoran Desert 1: 802. 1964.
Type locality: Mazatlán, Sinaloa, Mexico [Sonora].

malmeana Pax & Hoffm. Pflanzenreich 147, Addit. VII(Heft 85): 191. 1924; McVaugh, Bull. Torrey Bot.
Club 72: 281. 1945. (I. 1. A.)
Type locality: Cuyabá, Matto Grosso, Brazil.

marginata Chiov. Ann. Bot. Roma 18: 324. 1930; Fl. Somala 2: 391, fig. 224. 1932. (I. 2.)
Type locality: Somali Republic.

martiusii (Pohl) Baillon, Adansonia I. 4: 268. 1864; Muell. Arg. in DC. Prodr. 15(2): 1091. 1866; in Mart.
Fl. Brasil. 11(2): 494. 1874; Pax, Pflanzenreich 147 (Heft 42): 37. 1910; McVaugh, Bull. Torrey Bot.
Club 72: 277. 1945. (I. 4.)
Syn.: *Adenoropium martiusii* Pohl, Pl. Bras. Icon. Descr. 1:16. 1827.
Type locality: Malhada, Bahia, Brazil.

matacensis Castell. Bol. Acad. Nac. Ci. Cordoba 40: 255, fig. 6. 1958. (I. 1. A.)
Type locality: Matacos, Formosa, Argentina.

mcvaughii Dehgan & Webster, Madroño 25:36. 1978. (II. 1.)
Syn.: *J. curcas* var. *rufus* McVaugh, Bull. Torrey Bot. Club 72: 284. 1945.
Type locality: Playa Scandida, Jalisco, Mexico.

melanosperma Pax, Bot. Jahrb. Syst. 19: 110. 1895; Pflanzenreich 147 (Heft 42): 68. 1910; Hutchinson,
Fl. Trop. Afr. 6(1): 786. 1913; Andrews, Fl. Pl. Anglo-Egypt. Sudan 2: 83. 1952. (I. 5. A.)
Type locality: Jur Ghattas, Bahr el Ghazal, Sudan.

messinica E. A. Bruce: see **spicata**

microdonta Radcliffe-Smith, Kew Bull. 28: 284. 1973. (I. 5?)
Type locality: Masailand, Tanzania.

minor Urban, Symb. Ant. 9: 213. 1924; McVaugh, Bull. Torrey Bot. Club 72: 275. 1945; Alain, Fl. Cuba
3: 77. 1953. (I. 6. A.)
Type locality: Sierra de Nipe, Oriente, Cuba.

mollis Pax, Annuario Reale Ist. Bot. Roma 6: 184. 1896; Pflanzenreich 147 (Heft 42): 39. 1910; Hutchin-
son, Fl. Trop. Afr. 6(1): 796. 1913; Pax in Engler & Drude, Veg. Erde 9. III(2): 112. 1921; Radcliffe-
Smith, Kew Bull. 28: 521. 1973. (I. 1. B.)
Type locality: Tombe, Somali Republic [Kenya].

mollissima (Pohl) Baillon, Adansonia I. 4: 263. 1864; McVaugh, Bull. Torrey Bot. Club 72: 277. 1945.
(I. 4.)
var. **mollissima**
Syn.: *Adenoropium mollissimum* Pohl, Pl. Bras. Icon. Descr. 1:15. 1827.—*Jatropha pohliana* Muell.
Arg. Mém. Soc. Phys. Genève 17: 449. 1864; in DC. Prodr. 15(2): 1091. 1866; in Mart.
Fl. Brasil. 11(2): 492. 1874; Pax, Pflanzenreich 147 (Heft 42): 38. 1910.
Adenoropium luxurians Pohl, Pl. Bras. Icon. Descr. 1:16. 1827.—*Jatropha luxurians* (Pohl)
Baillon, Adansonia I. 4: 268. 1864.

---

5. The expanded description of *J. mahafalensis* by Jumelle (1920) strongly suggests that the plant may
be a species of *Givotia;* indeed, in the original description Jumelle and Perrier compared the plant to *Ri-
cinodendron,* a genus closely related to *Givotia.* The trilobed exstipulate leaves, the dioecious flower production, and the drupaceous fruit concur with the treatment of *G. madagascariensis* Baillon by Radcliffe-
Smith (Kew Bull. 22: 502. 1968). There are indeed some discrepancies between Jumelle's description and
the original one by Baillon (Bull. Soc. Linn. Paris 1: 810. 1889), especially the description by Jumelle of
the fruit as trilocular; according to Radcliffe-Smith, the fruits are always unilocular by abortion in *Givotia.*
However, whether or not Jumelle's plant is conspecific with *G. madagascariensis,* it is highly probable that
it belongs in *Givotia,* and there are thus no true species of *Jatropha* native to Madagascar.

Type locality: Malhada, Bahia, Brazil.
  var. **divergens** (Pohl) Muell. Arg. in DC. Prodr. 15(2): 1092. 1866.
    Syn.: *Adenoropium divergens* Pohl, Pl. Bras. Icon. Descr. 1: 15. 1827.—*Jatropha divergens* (Pohl)
      Baillon, Adansonia I. 4: 268. 1864.
    Type locality: Rio São Francisco, Minas Gerais, Brazil.
  var. **glabra** Muell. Arg. in DC. Prodr. 15(2): 1092. 1866.
    Type locality: Macara I., mouth of Orinoco River, Venezuela.
  var. **subglabra** Muell. Arg. in Mart. Fl. Brasil. 11(2): 494. 1874.
    Type locality: Pernambuco, Brazil.
  var. **velutina** Pax & Hoffm. Pflanzenreich 147, Addit. VII (Heft 85): 191. 1924.
    Type locality: Rio São Francisco, Bahia, Brazil.
  var. **villosa** (Pohl) Muell. Arg. in DC. Prodr. 15(2): 1091. 1866.
    Syn.: *Adenoropium villosum* Pohl, Pl. Bras. Icon. Descr. 1: 15. 1827.—*Jatropha villosa* (Pohl) Bail-
      lon, Adansonia I. 4: 268. 1864.
    Type locality: Malhada, Bahia, Brazil.
moluensis Sessé & Moc.: see **integerrima**
monroi S. Moore, J. Bot. 63: 147. 1925. (I. 5. A?)
  Type locality: Victoria, Rhodesia.
moranii Dehgan & Webster, Madroño 25:34. 1978 (II. 2.)
  Type locality: Cabo San Lucas, Baja California Sur, Mexico.
multifida L. Sp. Pl. 1: 1006. 1753; Salisb. Parad. Lond. t. 91. 1806; H.B.K. Nov. Gen. Sp. Pl. 2: 83. 1817;
  Griseb. Fl. Brit. W. Ind. 36. 1859; Muell. Arg. in DC. Prodr. 15(2): 1089. 1866; in Mart. Fl. Brasil. 11
  (2): 495, t. 69 fig. 1. 1874; Hook f. Fl. Brit. Ind. 5: 373. 1887; Pax in Engler, Pflanzenw. Ost-Afr. C:
  240. 1895; Urban, Symb. Ant. 4: 350. 1905; Pax, Pflanzenreich 147 (Heft 42): 40. 1910; Hutchinson,
  Fl. Trop. Afr. 6(1): 784. 1913; Fawc. & Rendle, Fl. Jam. 4(2): 313. 1920; Standley, Contr. U.S. Nat.
  Herb. 23: 637. 1923; McVaugh, Bull. Torrey Bot. Club 72: 277. 1945; Alain, Fl. Cuba 3: 76. 1953;
  Backer & Bakh. f. Fl. Java 1: 494. 1963; Gooding et al. Fl. Barbados 256. 1965; Airy Shaw, Kew Bull.
  26: 284. 1971; Little et al. Trees Puerto Rico & Virgin I. 2:420, t. 446. 1974. (I. 4.)
  Syn.: *Adenoropium multifidum* (L.) Pohl, Pl. Bras. Icon. Descr. 1: 16. 1827.
  Type locality: "America meridionali."
multiflora Pax & Hoffm.: see **macrocarpa**
mutabilis (Pohl) Baillon, Adansonia I. 4: 267. 1864; Muell. Arg. in DC. Prodr. 15(2): 1103. 1866; in Mart.
  Fl. Brasil. 11(2): 490. 1874; Pax, Pflanzenreich 147 (Heft 42): 83. 1910; McVaugh, Bull. Torrey Bot.
  Club 72: 276. 1945. (I?)
  Syn.: *Adenoropium mutabile* Pohl, Pl. Bras. Icon. Descr. 1: 14. 1827.
  Type locality: Bahia, Brazil.

nana Dalz. & Gibson, Bombay Fl. 229. 1861; Muell. Arg. in DC. Prodr. 15(2): 1083. 1866; Hook. f. Fl.
  Brit. Ind. 5: 382. 1887; Cooke, Fl. Pres. Bombay 94. 1906; Pax, Pflanzenreich 147 (Heft 42): 70. 1910.
  (I. 5. A?)
  Type locality: Poona, India.
natalensis Muell. Arg. Flora 47: 485. 1864; in DC. Prodr. 15(2): 1083. 1866; Wood, Natal Pl. 3: t. 242.
  1901; Pax, Pflanzenreich 147 (Heft 42): 65. 1910; Prain, Fl. Cap. 5(2): 422. 1920. (I. 5. A.)
  Type locality: Moor River Valley, Natal, South Africa.
neopauciflora Pax, Pflanzenreich 147 (Heft 42): 134. 1910; Standley, Contr. U.S. Nat. Herb. 23: 638.
  1923; McVaugh, Bull. Torrey Bot. Club 72: 287. 1945. (II. 3. C.)
  Syn.: *Mozinna pauciflora* Rose, Contr. U.S. Nat. Herb. 12: 282, t. 22. 1909.—*Jatropha pauciflora*
    (Rose) Pax, Pflanzenreich 147 (Heft 42): 82. 1910. (non *J. pauciflora* Griseb., 1865.)
    J. harmsiana Mattf. Fedde's Repert. 19: 121. 1923.
  Type locality: Tehuacán, Puebla, Mexico.
neriifolia Muell. Arg. Flora 47: 485. 1864; in DC. Prodr. 15(2): 1089. 1866; Pax, Pflanzenreich 147 (Heft
  42): 65. 1910; Hutchinson, Fl. Trop. Afr. 6(1): 781. 1913; Hutch. & Dalz. Fl. W. Trop. Afr. ed. 2, 1:
  397. 1958. (I. 5. A.)
  Type locality: Nupe, Nigeria.
nogalensis Chiov. Fl. Somala 1: 306, t. 35, fig. 2. 1929. (I. 3.)
  Type locality: Nogal Basin, Somali Republic.

nudicaulis Benth. Bot. Voy. Sulphur 165. 1846; Muell. Arg. in DC. Prodr. 15(2): 1092. 1866; Pax, Pflanzenreich 147 (Heft 42): 45. 1910; McVaugh, Bull. Torrey Bot. Club 72: 277. 1945. (I. 4.)
Type locality: Monte Christi, Colombia [Ecuador].

obbiadensis Chiov. Fl. Somala 1: 308, t. 36 fig. 2. 1929. (I. 5. A.)
Type locality: Obbia, Somali Republic.
oblanceolata Radcliffe-Smith, Kew Bull. 28: 522. 1973. (I. 3.)
Type locality: Meru Distr., Kenya.
officinalis Mart. ex Baillon: see **elliptica**
olivacea Muell. Arg.: see **ciliata**
opifera Mart.: see **elliptica**
orangeana Dinter ex P. Meyer, Mitt. Bot. Staatssaml. München 3: 612. 1960. (I?)
Type locality: Kahanstal, Southwest Africa.
ortegae Standley, Field Mus. Nat. Hist., Bot. Ser. 22: 38. 1940; McVaugh, Bull. Torrey Bot. Club 72: 288. 1945. (II. 3. A.)
Type locality: Sinaloa, Mexico.

pachypoda Pax, Pflanzenreich 147 (Heft 42): 47. 1910; McVaugh, Bull. Torrey Bot. Club 72: 277. 1945. (I?)
Type locality: Tarija, Bolivia.
palmatifida Baker, Kew Bull. 1895: 277. 1895; Pax Pflanzenreich 147 (Heft 42): 34. 1910; Hutchinson, Fl. Trop. Afr. 6(1): 787. 1913; Chiov. Fl. Somala 1: 305. 1929. (I. 1. B.)
Type locality: Golis Range, Somali Republic.
palmatifolia Ule, Bot. Jahrb. 42: 219. 1908; Pax, Pflanzenreich 147 (Heft 42): 31. 1910. (I?)
Type locality: Tambury, Bahia, Brazil.
palustris Sessé & Moc.: status dubious (see McVaugh, 1945)
pandurifolia Andr.: see **integerrima** var. **integerrima**
pandurifolia var. *coccinea* (Link) Pax: see **integerrima**
pandurifolia var. *latifolia* Pax: see **integerrima**
papyrifera Pax & Hoffm. Notizbl. Bot. Gart. Berlin 10: 385. 1928; McVaugh, Bull. Torrey Bot. Club 72: 277. 1945. (I. 6?)
Type locality: Villa Montes, Bolivia.
paradoxa (Chiov.) Chiov. Ann. Bot. Roma 18: 324. 1930. (I. 2.)
Syn.: *Collenucia paradoxa* Chiov. Fl. Somala 1: 177. 1929.
Type locality: Somali Republic.
parvifolia Chiov. Atti Soc. Nat. Mat. Modena 66: 18. 1935; Agnew, Upland Kenya Wild Flowers 218. 1974. (I?)
Type locality: Somali Republic [Kenya].
pauciflora Griseb.: see **integerrima**
pauciflora (Rose) Pax: see **neopauciflora**
paxii Croizat, J. Arnold Arb. 24: 168. 1943. (I. 6. A.)
Syn.: *J. flabellifolia* Pax & Hoffm. Pflanzenreich 147 (Heft 42): 52, fig. 20. 1910 (non [Pohl] Steud. 1840 = *Manihot esculenta*).
Type locality: Cuba.
pedatipartita O. Ktze. Rev. Gen. Pl. 3(2): 287. 1898; Pax, Pflanzenreich 147 (Heft 42): 30. 1910; McVaugh, Bull. Torrey Bot. Club 72: 281. 1945. (I. 1. A.)
Type locality: Parotani, Bolivia.
pedersenii Lourt. Ark. Bot. II. 3(5): 83. 1945; Allem, Rev. Bras. Biol. 37: 209. 1977. (I. 6. C; = **J. isabellii**?)
Type locality: Mburucuya, Corrientes, Argentina.
peiranoi Lourt. & O'Donn. Lilloa 9: 135, t. 14. 1943; McVaugh, Bull. Torrey Bot. Club 72: 280. 1945. (I. 1. A.)
Type locality: Belen, Catamarca, Argentina.
pelargoniifolia Courb. Ann. Sci. Nat. Bot. IV. 18: 150. 1862.[6] (I. 1. B.)

---

6. It is with considerable regret that we are adopting an obscure name for the common African-Arabian plant which has been called *J. glandulosa* Vahl or *J. villosa* (Forsk.) Muell. Arg. However, the latter name

Syn.: *Croton villosum* Forsk. Fl. Aegypt. Arab. 163. 1775.—*Jatropha villosa* (Forsk.) Muell. Arg. in
   DC. Prodr. 15(2): 1085. 1866 (non *J. villosa* Wight, 1848, nec (Pohl) Baillon, 1863).—*Aden-
   oropium forskalei* Pohl, Pl. Bras. Icon. Descr. 1: 15. 1827.
   Jatropha glandulosa Vahl, Symb. Bot. 1: 80. 1790 (nom. superfl.); Hutch. Fl. Trop. Afr. 6(1):
      793. 1913; Radcliffe-Smith, Kew Bull. 28: 521. 1973.—*J. villosa* var. *glandulosa* (Vahl) Pax,
      Pflanzenreich 147 (Heft 42): 45, fig. 17. 1910.
   J. villosa β *glabra* Muell. Arg. in DC. Prodr. 15(2): 1085. 1866.
   J. glandulosa var. *pelargoniifolia* (Courb.) Chiov. Fl. Somala 1: 306. 1929.
   J. glandulosa var. *sublobata* O. Schwartz, Mitt. Inst. Allg. Bot. Hamburg 10: 140. 1939.
   Type locality: "Ile de Dissée," Somali Republic [Arabia, Eritrea, Kenya].
peltata H.B.K.: see **humboldtiana**
peltata Sessé: status dubious (see McVaugh, 1945)
phillipseae Rendle, J. Bot. 36: 30. 1898; Pax, Pflanzenreich 147 (Heft 42): 37. 1910; Hutchinson, Fl.
   Trop. Afr. 6(1): 795. 1913. (I. 1. B.)
   Type locality: Wagga Mountains, Somali Republic.
plantanifolia Standley: see **malacophylla**
platyphylla Muell. Arg. in DC. Prodr. 15(2): 1077. 1866; Pax, Pflanzenreich 147 (Heft 42): 76. 1910;
   Standley, Contr. U.S. Nat. Herb. 23: 639. 1923; McVaugh, Bull. Torrey Bot. Club 72: 286. 1945. (II.
   2.)
   Type locality: Mexico [Sinaloa to Michoacán].
podagrica Hook. Bot. Mag. t. 4376. 1848; Muell. Arg. in DC. Prodr. 15(2): 1093. 1866; Urban, Symb. Ant.
   4: 350. 1905; Pax, Pflanzenreich 147 (Heft 42): 44. 1910; Fawc. & Rendle, Fl. Jam 4(2): 313. 1920;
   Standley, Contr. U.S. Nat. Herb. 23: 637. 1923; McVaugh, Bull. Torrey Bot. Club 72: 277. 1945; Standley & Steyermark, Fieldiana Bot. 24(6): 129. 1949; Backer & Bakh. f. Fl. Java 1: 494. 1963. (I. 4.)
   Type locality: "Central America" [Mexico].
pohliana Muell. Arg.: see **mollissima**
portoricensis Millsp.: see **hernandiifolia** var. **portoricensis**
prunifolia Pax in Engler, Pflanzenw. Ost.-Afr. C: 240. 1895; Bot. Jahrb. Syst. 23: 530. 1897; Pflanzenreich 147 (Heft 42): 54. 1910; Hutchinson, Fl. Trop. Afr. 6(1): 782. 1913. (I. 5. B.)
   Type locality: Usambara, Tanzania.
pseudo-curcas Muell. Arg. Linnaea 34: 208. 1865; in DC. Prodr. 15(2): 1080. 1866; Pax, Pflanzenreich
   147 (Heft 42): 77. 1910; Standley, Contr. U.S. Nat. Herb. 23: 642. 1923; McVaugh, Bull. Torrey Bot.
   Club 72: 284. 1945; Wilbur, J. Elisha Mitchell Sci. Soc. 70: 96. 1954. (II. 1.)
   Type locality: Oaxaca, Mexico.
pseudoglandulifera Pax: see **spicata**
puncticulata Pax & Hoffm. Pflanzenreich 147, Addit. VII (Heft 85): 192. 1924; McVaugh, Bull. Torrey
   Bot. Club 72: 280. 1945. (I. 1. A.)
   Type locality: Asunción, Paraguay.
purpurea Rose, Contr. U.S. Nat. Herb. 1: 357. 1895; Rose & Pax, Pflanzenreich 147 (Heft 42): 42, fig. 15.
   1910; Standley, Contr. U.S. Nat. Herb. 23: 637. 1923; McVaugh, Bull. Torrey Bot. Club 72: 280. 1945;
   Shreve & Wiggins, Veg. Fl. Sonoran Desert 1: 801. 1964. (I. 1?)[7]
   Type locality: Agiabampo, Sonora, Mexico [Sinaloa].

ribifolia (Pohl) Baillon, Adansonia I. 4: 286. 1864; Pax, Pflanzenreich 147 (Heft 42): 28. 1910; McVaugh,
   Bull. Torrey Bot. Club 72: 280. 1945. (I. 1. A.)
   var. **ribifolia**
      Syn.: *Adenoropium ribifolium* Pohl, Pl. Bras. Icon. Descr. 1: 15. 1827.—*Jatropha gossypifolia* α
         ribifolia (Pohl) Muell. Arg. in Mart. Fl. Brasil. 11(2): 491. 1874.

---

is a later homonym of names given by Wight (1848) and Baillon (1863). Vahl's name *J. glandulosa* appears
to be illegitimate because it was superfluous when published; he should have adopted the epithet *villosa*
when he transferred Forskahl's species from *Croton*. *Adenoropium forskalei* of Pohl is likewise a superfluous name, since it was based directly on Forskahl's species, and he still could have adopted the epithet
villosa in 1827. The first legitimate name, therefore, appears to be that of Courbet in 1862, based on a different type.

7. *Jatropha purpurea* appears to have been correctly placed by Pax in sect. *Jatropha;* it is isolated sys-
tematically from other North American species, and further study will probably justify Pax's creation of
a special subsection for it.

Type locality: Bahia, Brazil.
> var. **ambigua** Pax, Pflanzenreich 147 (Heft 42): 28. 1910.
>> Type locality: Olimpo, Paraguay.

ribifolia var. *breviloba* Pax: see **breviloba**

ricinifolia Pax, Pflanzenreich 147 (Heft 42): 34, fig. 11. 1910. (I. 1. A?)
> Type locality: Chaco, Paraguay.

rigidifolia Pax & Hoffm. Pflanzenreich 147, Addit. V (Heft 63): 398. 1914; McVaugh, Bull. Torrey Bot. Club 72: 280. 1945; Allem, Rev. Bras. Biol. 37: 209. 1977. (I. 1. A; = **J. isabellii?**)
> var. **rigidifolia**
>> Type locality: Alto Paraná, Paraguay.
> var. **glabrescens** Pax & Hoffm. Pflanzenreich 147, Addit VII (Heft 85): 192. 1924.
>> Type locality: Cruz Alta, Río Grande do Sul, Brazil.

riojae Miranda, Anales Inst. Biol. Univ. Nac. Mex. 13: 456. 1942; McVaugh, Bull. Torrey Bot. Club 72: 291. 1945. (II. 3. A.)
> Type locality: Acatlán, Puebla, Mexico.

rivae Pax, Annuario Reale Ist. Bot. Roma 6: 185. 1896; Pflanzenreich 147 (Heft 42): 57, fig. 21. 1910; Hutchinson, Fl. Trop. Afr. 6(1): 799. 1913. (I. 3.)
> Type locality: Dolo, Somali Republic.

robecchii Pax, Annuario Reale Ist. Bot. Roma 6: 184. 1896; Pflanzenreich 147. (Heft 42): 83. 1910; Hutchinson, Fl. Trop. Afr. 6(1): 779. 1913. (I. 5?)
> Type locality: Somali Republic.

robertii S. Moore, J. Bot. 56: 212. 1918; McVaugh, Bull. Torrey Bot. Club 72: 277. 1945. (I. 6. A?)
> Type locality: Corumbá, Matto Grosso, Brazil.

rufescens Brandegee, Univ. Cal. Publ. Bot. 4: 88. 1910; Pax, Pflanzenreich 147, Addit. V (Heft 63): 399. 1914; Standley, Contr. U.S. Nat. Herb. 23: 642. 1923; McVaugh, Bull. Torrey Bot. Club 72: 287. 1945. (II?)
> Type locality: Tlaciulotepec, Puebla, Mexico.

sabdariffa Schweinf.: see **aethiopica**

schlechteri Pax, Bot. Jahrb. Syst. 28: 24. 1899; Pflanzenreich 147 (Heft 42): 67. 1910; Prain, Fl. Cap. 5(2): 427. 1920. (I. 5. A.)
> Type locality: Koomati Poort, Transvaal, South Africa.

schweinfurthii Pax, Bot. Jahrb. Syst. 19: 110. 1894; Pflanzenreich 147 (Heft 42): 70. 1910; Hutchinson, Fl. Trop. Afr. 6(1): 794. 1913; Andrews, Fl. Pl. Anglo-Egypt. Sudan 2: 85. 1952; Radcliffe-Smith, Kew Bull. 28: 285. 1973. (I. 5?)
> ssp. **schweinfurthii**
>> Type locality: Sudan.
> ssp. **atrichocarpa** Radcliffe-Smith, Kew Bull. 28: 285. 1973.
>> Type locality: Ruaha Nat. Park, Tanzania [Rhodesia].

seineri Pax, Bot. Jahrb. Syst. 43: 84. 1909; Pflanzenreich 147 (Heft 42): 65. 1910; Hutchinson, Fl. Trop. Afr. 6(1): 781. 1913. (I. 5. A.)
> Syn.: *J. humilis* N. E. Brown, Kew Bull. 1909: 129. 1909.
> Type locality: Caprivi Peak, Southwest Africa [Botswana]

setifera Hutchinson, Botanist in So. Afr. 397. 1946 (I. 5. A?)
> Type locality: Doornpoort, Transvaal, South Africa.

somalensis Pax, Pflanzenreich 147 (Heft 42): 68. 1910; Hutchinson, Fl. Trop. Afr. 6(1): 786. 1913. (I. 5. A.)
> Type locality: Djuba, Somali Republic.

spathulata (Ortega) Muell. Arg.: see **dioica** var. **dioica**.

spathulata var. *sessiliflora* (Hook.) Muell. Arg.: see **dioica** var. **dioica**

spicata Pax, Bot. Jahrb. Syst. 19: 109. 1894; in Engler, Pflanzenw. Ost-Afr. C: 240. 1895; Pflanzenreich 147 (Heft 42): 36. 1910; Hutchinson, Fl. Trop. Afr. 6(1): 790. 1913; Pax in Engler & Drude, Veg. Erde 9. III(2): 112. 1921; Dale & Greenway, Kenya Trees & Shrubs 206. 1961; Radcliffe-Smith, Kew Bull. 28: 285. 1973. (I. 1. B.)
> var. **spicata**
>> Syn.: *J. pseudoglandulifera* Pax, Pflanzenreich 147 (Heft 42): 34. 1910.
>> *J. kilimanscharica* Pax & Hoffm. Pflanzenreich 147 (Heft 42): 40. 1910.

J. messinica E. A. Bruce, Bothalia 6: 226. 1951.

Type locality: Teita, N'dara, Kenya [Angola, Transvaal].

var. **zanguebarica** Hutchinson, Fl. Trop. Afr. 6(1): 1053. 1913; Pax & Hoffm. Pflanzenreich 147, Addit. V: 398. 1914.

Type locality: Lamu Island.

spinosa Vahl, Symb. Bot. 1: 79. 1790; Krauss, Bot. Jahrb. Syst. 35: 720. 1905; Pax, Pflanzenreich 147 (Heft 42): 55. 1910; Blatter, Rec. Bot. Surv. India 8: 435. 1923. (I. 3.)

var. **spinosa**

Syn.: *Croton spinosum* sensu Forsk. Fl. Aegypt. Arab. 163. 1775 (non L. 1753).

J. aculeata F. G. Dietr. Lexicon Gaertn. Bot. Nachtr. 4: 76. 1818; Muell. Arg. in DC. Prodr. 15(2): 1083. 1866.

Adenoropium spinosum Pohl, Pl. Bras. Icon. Descr. 1: 15. 1827.

Type locality: Uahfad, Arabia.

var. **crenulata** Pax, Pflanzenreich 147 (Heft 42): 56. 1910; Swartz, Inst. Allg. Bot. Hamburg 10: 140. 1939.

Type locality: southern Arabia.

var. **somalensis** Pax, Pflanzenreich 147 (Heft 42): 56. 1910; Hutchinson, Fl. Trop. Afr. 6(1): 780. 1913.

Type locality: Somadu, Somali Republic.

standleyi Steyermark, Publ. Field Mus. Nat. Hist., Bot. Ser. 22: 152. 1940; McVaugh, Bull. Torrey Bot. Club 72: 291. 1945. (II. 3. A.)

Type locality: Tehuantepec, Oaxaca, Mexico.

staphysagrifolia Miller: see **gossypiifolia** var. **gossypiifolia**

stigmatosa Pax & Hoffm. Pflanzenreich 147, Addit. VII (Heft 85): 192. 1924; McVaugh, Bull. Torrey Bot. Club 72: 291. 1945. (I?)

Type locality: Matto Grosso, Brazil.

stuhlmanii Pax in Engler, Pflanzenw. Ost-Afr. C: 240. 1895; Bot. Jahrb. Syst. 23: 530. 1897; Pflanzenreich 147 (Heft 42): 39, fig. 12. 1910; Hutchinson, Fl. Trop. Afr. 6(1): 796. 1913; Pax in Engler & Drude, Veg. Erde 9.III(2): 112. 1921; Radcliffe-Smith, Kew Bull. 28: 286. 1973. (I. 1. B.)

Syn.: *J. batawe* Pax, Bot. Jahrb. Syst. 28: 420. 1900; Pflanzenreich 147 (Heft 42): 40. 1910.

Type locality: Usaramo, Tanzania.

sympetala Blake & Standley, Proc. Biol. Soc. Wash. 33: 118. 1920; Standley, Contr. U.S. Nat. Herb. 23: 639. 1923; McVaugh, Bull. Torrey Bot. Club 72: 291. 1945. (II. 3. C.)

Type locality: Playa de Coyula, Oaxaca, Mexico.

tacumbensis Pax & Hoffm. Pflanzenreich 147, Addit. VII (Heft 85): 191. 1924; McVaugh, Bull. Torrey Bot. Club 72: 280. 1945. (I. 1. A.)

Type locality: Cerro Tacumba, Asunción, Paraguay.

tanjorensis Ellis & Saroja, J. Bombay Nat. Hist. Soc. 58: 834. 1962. (I. 1. A.)

Type locality: Tanjore, Madras, India.

tetracantha Chiov. Fl. Somala 2: 393. 1932. (I?)

Type locality: Somali Republic.

thyrsantha Pax & Hoffm. Pflanzenreich 147, Addit. V (Heft 63): 397. 1914; McVaugh, Bull. Torrey Bot. Club 72: 280. 1945. (I. 1. A.)

Type locality: Charagua, Bolivia.

transiens Pax, Pflanzenreich 147 (Heft 42): 73. 1910: = *J. isabellii* X *dissecta?* (see Lourteig & O'Donnell, Lilloa 9: 140. 1943; Allem, Rev. Bras. Biol. 37: 212. 1977.)

trifida Chiov. Fl. Somala 2: 388. 1932. (I. 1. B.)

Syn.: *J. brockmanii* var. *leiosepala* Chiov. Result. Scient. Miss. Stefanini-Paoli (Publ. Istit. Stud. Sup. Firenze) 1: 162. 1916; Pax & Hoffm. Pflanzenreich 147, Addit. VII (Heft 85): 191. 1924.

Type locality: Somali Republic.

tropaeolifolia Pax, Pflanzenreich 147 (Heft 42): 56. 1910; Hutchinson, Fl. Trop. Afr. 6(1): 789. 1913. (I. 3.)

Type locality: Merehan, Somali Republic.

tuberosa Pax, Bot. Jahrb. Syst. 19: 111. 1894; Pflanzenreich 147 (Heft 42): 68, fig. 27. 1910; Hutchinson, Fl. Trop. Afr. 6(1): 785. 1913; Pax in Engler & Drude, Veg. Erde 9. III(2): 116. 1921; Andrews, Fl. Pl. Anglo-Egypt. Sudan 2: 83. 1952; Radcliffe-Smith, Kew Bull. 28: 286. 1973. (I. 5. A.)

Type locality: Mangob, Bahr el Ghazal, Sudan [Uganda].

tupifolia Griseb. Abh. Konigl. Ges. Wiss. Göttingen 12: 170. 1865; Muell. Arg. in DC. Prodr. 15(2): 1094.
1866; Pax, Pflanzenreich 147 (Heft 42): 52. 1910; McVaugh, Bull. Torrey Bot. Club 72: 275. 1945;
Alain, Fl. Cuba 3: 77. 1953. (I. 6. A.)
Syn.: *J. diversifolia β tupifolia* (Griseb.) Gomez, Anal. Hist. Nat. Madrid 23: 51. 1894.
Type locality: eastern Cuba.

unicostata Balf. f. Proc. Roy. Soc. Edinb. 12: 94. 1884; Trans. Roy. Soc. Edinb. 31: 272, t. 90. 1888; Pax,
Pflanzenreich 147 (Heft 42): 53. 1910. (I. 5. A.)
Type locality: Socotra.

variegata Vahl, Symb. Bot. 1: 79, t. 21. 1790; Muell. Arg. in DC. Prodr. 15(2): 1084. 1866; Pax, Pflanzen-
reich 147 (Heft 42): 54. 1910; Blatter, Rec. Bot. Surv. India 8: 436. 1923; Schwartz, Mitt. Inst. Allg.
Bot. Hamburg 10: 140. 1939. (I. 5.?)
Syn.: *Croton variegatus* sensu Forsk. Fl. Aegypt. Arab. 163. 1775 (non L. 1753).
Adenoropium variegatum Pohl, Pl. Bras. Icon. Descr. 1: 14. 1827.
Type locality: Zebid, Yemen.

variifolia Pax, Pflanzenreich 147 (Heft 42): 54. 1910; Prain, Fl. Cap. 5(2): 420. 1920. (I. 5. B.)
Syn.: *J. heterophylla* Pax, Bot. Jahrb. Syst. 28: 1899 (non Steud. 1840 = *Manihot heterophylla*).
Type locality: Koomati Poort, Transvaal, South Africa.

velutina Pax & Hoffm. Pflanzenreich 147 (Heft 42): 37. 1910. (I. 1. B.)
Syn.: *J. acerifolia* Pax, Bot. Jahrb. Syst. 19: 109. 1894 (non Salisb. 1796); in Engler, Pflanzenw. Ost-
Afr. C: 240. 1895; Pflanzenreich 147 (Heft 42): 37. 1910; Hutchinson, Fl. Trop. Afr. 6(1):
797. 1913; Pax in Engler & Drude, Veg. Erde 9.III(2): 112. 1921; Radcliffe-Smith, Kew Bull.
28: 283. 1973.
Type locality: Taro, Kenya.

vernicosa Brandegee, Zoe 5: 206. 1905; Pax, Pflanzenreich 147 (Heft 42): 85. 1910; Standley, Contr. U.S.
Nat. Herb. 23: 638. 1923; McVaugh, Bull. Torrey Bot. Club 72: 288. 1945; Shreve & Wiggins, Veg. Fl.
Sonoran Desert 1: 803. 1964. (II. 3. A.)
Type locality: Cape Region, Baja California Sur, Mexico.

villosa Wight, Icon. Pl. Ind. Or. 4: t. 1169. 1848; Balakr. Bull. Bot. Surv. India 3: 40. 1962. (II. 1.)
Syn.: *J. wightiana* Muell. Arg. in DC. Prodr. 15(2): 1080. 1866; Hook. f. Fl. Brit. Ind. 5: 383. 1887;
Pax, Pflanzenreich 147 (Heft 42): 80. 1910.
Type locality: Coimbatore, India.

villosa (Pohl) Baillon: see **mollissima** var.**villosa**

villosa (Forsk.) Muell. Arg.: see **pelargoniifolia**

villosa var. *glandulosa* (Vahl) Pax: see **pelargoniifolia**

weberbaueri Pax & Hoffm. Pflanzenreich 147 (Heft 42): 45, fig. 16. 1910; McVaugh, Bull. Torrey Bot.
Club 72: 277. 1945; Macbride, Field Mus. Nat. Hist., Bot. Ser. 13(3A): 162. 1951. (I. 4.)
Type locality: Luya, Marañón Valley, Peru [Educador].

weddelliana Baillon, Adansonia I. 4: 267. 1864; Muell. Arg. in DC. Prodr. 15(2): 1090. 1866; in Mart. Fl.
Brasil. 11(2): 494. 1874; Chodat & Hassler, Bull. Herb. Boiss. II. 5: 631. 1905; Pax, Pflanzenreich 147
(Heft 42): 36. 1910; McVaugh, Bull. Torrey Bot. Club 72: 277. 1945; Lourteig, Ark. Bot. II. 3(5): 84.
1954; Allem, Rev. Bras. Biol. 37: 217. 1977. (I. 4.)
Type locality: Paraguay [Matto Grosso, Brazil].

wightiana Muell. Arg.: see **villosa** Wight

woodii O. Ktze. Rev. Gen. Pl. 3(2): 287. 1898; Pax, Pflanzenreich 147 (Heft 42): 66, fig. 26. 1910; Prain,
Fl. Cap. 5(2): 425. 1920. (I. 5. B.)
Syn.: *J. woodii* var. *vestita* Pax, Bot. Jahrb. Syst. 43:84. 1909.
J. woodii var. *kuntzei* Pax, Pflanzenreich 147 (Heft 42): 66. 1910.
Type locality: Ladysmith, Natal, South Africa.

yucatanensis Briq.: see **curcas**

zeyheri Sond. Linnaea 23: 117. 1850; Muell. Arg. in DC. Prodr. 15(2): 1088. 1866; Pax, Pflanzenreich
147 (Heft 42): 68. 1910; Prain, Fl. Cap. 5(2): 426. 1920. (I. 5. A.)
var. **zeyheri**

Type locality: Cape of Good Hope, South Africa.
var. **platyphylla** Pax, Pflanzenreich 147 (Heft 42): 68. 1910.
   Syn.: *J. brachyadenia* Pax & Hoffm. Pflanzenreich 147 (Heft 42): 66. 1910.
   Type locality: Boshveld, Transvaal, South Africa.
var. **subsimplex** Prain, Fl. Cap. 5(2): 426. 1920.
   Type locality: Shilouvane, Transvaal, South Africa.

Literature Cited

Adanson, M.
1763   Familles des plantes XLV. Famille les Titimales. Tithymali. pp. 346-358.

Alain, H.
1953.   *Jatropha. In* H. Leon (ed.), Flora de Cuba 3:75-77.

Allard, K. W.
1965.   Genetic systems associated with predominantly self-pollinated species, pp. 49-75.
In H. G. Baker & G. L. Stebbins (eds.), Genetics of colonizing species. Academic
Press, N.Y.

Allem, A. C.
1977.   Notas sistemáticas y nuevos sinónimos en Euphorbiaceae de América del Sur-IV.
Rev. Brasil. Biol. 37:209-221.

Arnoldi, W.
1912.   Zur Embryologie einiger Euphorbiaceen. Trudy Bot. Muz. Imp. Akad. Nauk. 9:
136-154.

Bailey, I. W.
1956.   Nodal anatomy and vasculature of seedlings. J. Arnold Arb. 37:269-287.

Bailey, I. W., and C. G. Nast.
1945.   The comparative morphology of the Winteraceae. VII. Summary and conclusions.
J. Arnold Arb. 26:37-47.

Beille, M. L.
1902.   Recherches sur le développement floral des Disciflores. Actes Soc. Linn. Bordeaux
61:231-410.

Baillon, H.
1858.   Étude générale du groupe de Euphorbiacées, pp. 294-296.

Bernhard, F.
1966.   Contribution à l'étude de glandes foliares chez les Crotonoidées. Mém. Inst. Franc.
Afrique Noire 75:67-156.

Cannon, W. A.
1911.   The root habits of desert plants. Carnegie Institute, Washington, Publ. 131. 96pp.

Capinpin, J. M., and V. C. Bruce
1955.   Floral biology and cytology of *Manihot utiliissima.* Philipp. Agric. 39:306-316.

Cavanilles, A. J .
1799.   Icones et descriptiones plantarum. 5:17, pl. 429.

Chiovenda, E.
1929.   Flora Somala, vol. 1. Publ. Ministr. Colonie, Roma 7:305-309.

Cooke, T.
1906.   *Jatropha,* pp. 93-95. *In* The flora of the Presidency of Bombay. Printed by P. C.
Ray, Sri Gouranga Press, Calcutta.

Corner, E. J. H.
1976. The seeds of Dicotyledons. Vols. I & II. Cambridge University Press, Cambridge.

Datta, N.
1967. *In* IOPB chromosome number reports XII. Taxon 16:341-350.
Dehgan, B.
1976. Experimental and evolutionary studies of relationships in the genus *Jatropha* L. (Euphorbiaceae). PhD. Dissertation, Dept. of Botany, Univ. Calif. Davis. 436pp.
1978. Comparative anatomy of the petiole and infrageneric relationships in *Jatropha* (Euphorbiaceae). Bot. Gazette, in press.
Dehgan, B., and M. C. Craig.
1978. Types of laticifers and crystals in *Jatropha* and their taxonomic implications. Amer. J. Bot. 65:345-352.
Dehgan, B., and G. L. Webster.
1978. Three new species of *Jatropha* (Euphorbiaceae) from Western Mexico. Madroño 25:30-39.
Dilcher, D. L.
1974. Approaches to the identification of angiosperm leaf remains. Bot. Rev. 40:2-157.
Dillenius, J. J.
1732. Hortus Elthamensis. London, 2 vols, 325 pl.
Duke, J. A.
1965. Keys for the identification of seedlings of some prominent woody species in eight forest types in Puerto Rico. Ann. Missouri Bot. Gard. 52:314-350.
1969. On tropical tree seedlings. I. Seeds, seedlings, systems and systematics. Ann. Missouri Bot. Gard. 56:125-161.
Eames, A. J.
1962. Morphology of angiosperms. McGraw-Hill, New York.
Ehrendorfer, F.
1973. Adaptive significance of major taxonomic characters and morphological trends in angiosperms, pp. 317-327. *In* V. H. Hywood (ed.), Taxonomy and ecology. The Systematic Association Special Volume No. 5. Academic Press, London.
Ellis, J. L. and T. L. Saroja.
1961. A new species of *Jatropha* from South India. J. Bombay Nat. Hist. Soc. 58:834-836.
Erdtman, G.
1952. Pollen morphology and plant taxonomy: Angiosperms. Almqvist and Wiksell, Stockholm. 539pp.
Esau, K.
1940. Developmental anatomy of the fleshy storage organ of *Daucus carota*. Hilgardia 13:175-226.
1965. Plant anatomy, 2nd ed. John Wiley & Sons, Inc. New York. 767pp.

Faegri, K., and L. Van der Pijl.
1970. The principles of pollination ecology. Pergamon Press, New York. 291pp.
Fekete, G., and J. Sz. Lacza.
1970. A survey of plant life-form systems and the respective research approaches. II. Ann. Hist.-Nat. Mus. Natl. Hung. 62:115-127.
1971. A survey of plant life-form systems and the respective research approaches. III. Ann. Hist.-Nat. Mus. Natl. Hung. 63:37-50.
Fernald, M. L.
1950. Gray's manual of botany, 8th ed. American Book Co., New York. xiv+1632pp.
Fosberg, F. R.
1976. Revision in the flora of St. Croix, U.S. Virgin Islands. Rhodora 78:79-119.

Foster, A. S.
   1938.   Structure and the growth of the shoot apex in *Ginkgo biloba*. Bull. Torrey Bot.
           Club 65:531-556.
Foster, A. S., and E. M. Gifford, Jr.
   1974.   Comparative morphology of vascular plants, 2nd ed., W. H. Freeman, and Co., San
           Francisco. 751pp.
Froembling, W.
   1896.   Anatomisch-systematische Untersuchung von Blatt und Axe der Crotoneen und
           Euphyllantheen. Inaug-diss., Cassel. 76pp.

Galun, E., Y. Jung, and A. Lang.
   1963.   Morphogenesis of floral buds of cucumber cultured in vitro. Developm. Biol. 6:
           370-387.
Gaucher, L.
   1902.   Recherches anatomiques sur les Euphorbiacées. Ann. Sci. Nat. Bot. VIII. 15:161-
           309.
Gifford, E. M. Jr., and G. E. Corson, Jr.
   1971.   The shoot apex in seed plants. Bot. Rev. 37:143-229.
Gram, B.
   1895-96.Om froskallen bygning has Euphorbiaceerne. Bot. Tidsskr. 20:358-389.
Grant, V.
   1971.   Plant speciation. Columbia University Press, New York. 435pp.
   1975.   Genetics of flowering plants. Columbia University Press, New York. 514pp.
Grisebach, A. H. R.
   1859.   Flora of the British West Indian Islands, p. 36. Wheldon & Wesley, Ltd. and
           Hafner Publ. Co., Codicot, Herts., New York.

Hallé, F.
   1971.   Architecture and growth of tropical trees exemplified by the Euphorbiaceae. Bio-
           tropica 3:56-62.
Hans, A. S.
   1973.   Chromosomal conspectus of Euphorbiaceae. Taxon 22:591-636.
Heslop-Harrison, J.
   1959.   Growth substances and flower morphogenesis. J. Linn. Soc., Bot. 56:269-281.
   1964.   Sex expression in flowering plants. Brookhaven Symp. Biol. 16:109-125.
Hickey, L. J.
   1973.   Classification of the architecture of dicotyledonous leaves. Amer. J. Bot. 60:17-
           33.
Hickman, J. C.
   1974.   Pollination by ants: a low energy system. Science 184:1290-1292.
Hitchcock, A. S., and M. L. Green.
   1935.   International rules of botanical nomenclature, 3rd. ed., pp. 139-143.
Hocking, B.
   1975.   Ant-plant mutualism: evolution and energy, pp. 78-90. *In* L. E. Gilbert and P.
           H. Raven (eds.), Symposium V. First International Congress of Systematic and
           Evolutionary Biology, Boulder, Colorado, August, 1973. University of Texas
           Press, Austin.
Holm, T.
   1899.   Seedlings of *Jatropha multifida* L. and *Persea gratissima* Gartn. Bot. Gaz. 28:60-
           64.
Howard, R. A.
   1973.   The vegetation of the Antilles, pp. 1-38. *In* A. Graham (ed.), Vegetation and vege-
           tational history of northern Latin America. Elsevier Scientific Publ. Co., Amster-
           dam.

Hutchinson, J.
1913.　Nyctaginaceae to Euphorbiaceae. *In* P. Thiselton-Dyer (ed.), Flora of tropical Africa 6(1):775-798. L. Reeve & Co., London.

Ingram, J.
1957.　Notes on the cultivated Euphorbiaceae: 1. the flowers of the Euphorbiaceae; 2. *Cnidoscolus* and *Jatropha*. Baileya 5:107-117.

International Code of Botanical Nomenclature.
1972.　Adopted by the 11th Intern. Bot. Congr., Seattle, Aug. 1969. Prepared and edited by F. A. Stafleu, et al. A. Oesthoeke's Uitgeversmaatschappij N. V. Utrecht—Netherlands, for the Intern. Assoc. Plant Tax.

Jussieu, A. L.
1789.　Genera plantarum, Paris [Facsimile edition Weinheim, 1964].

Kajale, L. B., and G. V. Rao.
1943.　Pollen and embryo sac of two Euphorbiaceae. J. Indian Bot. Soc. 12:229-236.

Kakkar, L., and G. S. Paliwal.
1972.　Foliar venation and laticifers in *Jatropha gossypifolia*. Beitr. Biol. Pflanzen 48: 425-432.

Koriba, K.
1958.　On the periodicity of tree-growth in the tropics, with the reference to the mode of branching, the leaf-fall, and the formation of the resting bud. Gard. Bull. Straits Settlem. 17:11-81.

Lacza, J. Sz., and G. Fekete.
1969.　A survey of plant life-form systems and the respective research approaches. I. Ann. Hist.-Nat. Mus. Natl. Hung. 61:129-139.
1972.　A survey of plant life-forms systems and the respective research approaches IV. Ann. Hist.-Nat. Mus. Natl. Hung. 64:53-62.

Leppik, E. E.
1969.　Morphogenic classification of flower types. Phytomorphology 18:451-466.

Linnaeus, C.
1737.　Genera Plantarum, 1st. ed. [*Jatropha*, p. 288].
1738.　Hortus Cliffortianus [p. 445].
1753.　Species Plantarum [*Jatropha*, pp. 1006-1007].
1763.　Species Plantarum, 2nd ed. [*Jatropha*, pp. 1428-1430].
1764.　Genera Plantarum, 6th ed. [No. 1084, p. 503].

Lourteig, A., and C. A. O'Donnell.
1943.　Euphorbiaceae Argentinae: *Jatropha*. Lilloa 9:118-139.

Lynch, S. P., and G. L. Webster.
1976.　A new technique of preparing pollen for scanning electron microscopy. Grana 15: 127-136.

MacKenzie, K. K.
1929.　Type of the genus *Jatropha*. Bull. Torrey Bot. Club 56:213-215.

Mandl, L.
1926.　Beitrage zur Kenntnis der Anatomie der Samen mehrerer Euphorbiaceen-Arten. Oesterr. Bot. Z. 75:1-17.

McVaugh, R.
1944.　The genus *Cnidoscolus:* generic limits and intergeneric groups. Bull. Torrey Bot. Club 71:457-474.
1945a.　The Jatrophas of Cervantes and of the Sessé and Moçiño Herbarium. Bull. Torrev Bot. Club 72:31-41.

1945b.   The genus *Jatropha* in America: principal intrageneric groups. Bull. Torrey Bot. Club 72:271-294.

Mehra, P. N., and A. S. Hans.
   1969.   *In* IOPB chromosome number reports XXI. Taxon 18:310-315.

Metcalfe, C. R., and L. Chalk.
   1950.   Anatomy of dicotyeldons. II:725-1500. Oxford.

Michaelis, P.
   1924.   Blutenmorphologische Untersuchungen an den Euphorbiaceen. Bot. Abh. 3:1-150.

Miller, P.
   1754.   The gardeners dictionary, 4th ed. London.

Miller, K. I., and G. L. Webster.
   1962.   Systematic position of *Cnidoscolus* and *Jatropha*. Brittonia 14:174-180.

Mueller Argoviensis, J.
   1866.   *Jatropha. In* De Candolle (ed.) Prodromus Systematis Naturalis Regni Vegetabilis 15(2):1076-1105.
   1874.   *Jatropha. In* Martius (ed.), Flora Brasilensis 11(2):486-502.

Muller, J.
   1973.   Pollen morphology of *Barringtonia calyptrocalyx* K. Sch. (Lecythidaceae). Grana Palynol. 13:29-44.

Nair, N. C., and V. Abraham.
   1962.   Floral morphology of a few species of Euphorbiaceae. Proc. Indian Acad. Sci. 56:1-12.

Nanda, P. C.
   1962.   Chromosome number of some trees and shrubs. J. Indian Bot. Soc. 41:266-268.

Ortega, C. G.
   1799.   Novarum, aut rariorum plantarum hort. veg. botan. matri. descriptionum decades. p. 105, pl. 13.

Pax, F.
   1910.   Euphorbiaceae–Jatropheae. *In* A. Engler (ed.) Das Pflanzenreich IV. 147(Heft 42):1-148. Verlag von Wilhelm Englemann, Leipzig.

Pax, F., and K. Hoffmann.
   1919.   Euphorbiaceae. *In* A. Engler (ed.), Das Pflanzenreich IV. 147. Addit. VI(Heft 68): 38. Verlag von Wilhelm Englemann, Leipzig.
   1931.   Euphorbiaceae–Jatrophinae. *In* A. Engler and K. Prantl (eds.), Die Nat. Pflanzenfam. ed. 2, 19c: 160.

Perceval, M. S.
   1965.   Floral biology. Pergamon Press Ltd., Oxford. 243pp.

Perry, B. A.
   1943.   Chromosome number and phylogenetic relationships in the Euphorbiaceae. Amer. J. Bot. 30:527-543.

Pohl, J. E.
   1827.   Plantarum Brasiliae Icones et Descriptiones Vindobonae 1:12, t. 9.

Prain, D.
   1920.   Euphorbiaceae; *Jatropha. In* P. Thiselton-Dyer (ed.), Flora Capensis 5(2):418-427. L. Reeve & Co., London.

Punt, W.
   1962.   Pollen morphology of Euphorbiaceae with special reference to taxonomy. Wentia 7:1-116.

Radford, A. E., W. C. Dickison, J. R. Massey, and C. R. Bell.
   1974.   Vascular plant systematics. Harper & Row Publishers, New York. 891pp.

Rao, C. V., and T. Ramalakshmi.
  1968.   Floral anatomy of Euphorbiaceae. I. Some non-cyathium taxa. J. Indian Bot. Soc. 47:278-300.
Rao, V.
  1971.   Anatomy of the inflorescence of some Euphorbiaceae with a discussion of phylogeny and evolution of the inflorescence including the cyathium. Bot. Not. 124: 39-64.
Raunkaier, C.
  1934.   The life forms of plants and statistical plant geography. Clarendon Press, Oxford. xvi+632pp.
Rickett, H. W.
  1944.   The classification of inflorescences. Bot. Rev. 10:187-231.
  1955.   Materials for a dictionary of botanical terms. III. Inflorescences. Bull. Torrey Bot. Club 82:419-445.
Rogers, D. J., and S. G. Appan.
  1973.   *Manihot, Manihotoides* (Euphorbiaceae). Flora Neotropica Monogr. 13:1-272.
Rupert, E. A., B. Dehgan, and G. L. Webster.
  1970.   Experimental studies of relationships in the genus *Jatropha*. I. *J. curcas* X *integerrima*. Bull. Torrey Bot. Club 97:321-325.

Sachs, R. M.
  1965.   Stem elongation. Annual Rev. Pl. Physiol. 16:73-96.
Sarmiento, G.
  1976.   Evolution of arid vegetation in tropical America, pp. 65-99. *In* D. W. Goodall (ed.), Evolution of desert biota. Univ. of Texas Press, Austin.
Sehgal, L., and G. S. Paliwal.
  1974.   Studies on the leaf anatomy of *Euphorbia*. II. Venation patterns. J. Linn. Soc., Bot. 68:173-208.
Seifriz, W.
  1943.   The plant life in Cuba. Ecol. Monogr. 13:375-426.
Shreve, F., and I. L. Wiggins.
  1964.   Vegetation and flora of the Sonoran Desert. Stanford University Press, Stanford, California. Vol. I:x+840pp.
Singh, R. P.
  1970.   Structure and development of seeds in Euphorbiaceae: *Jatropha* species. Beitr. Biol. Pflanzen 47:79-90.
Small, J. K.
  1933.   Manual of Southeastern flora. Univ. of North Carolina Press, Chapel Hill, N. C., pp. 791.
Snow, R.
  1963.   Alcoholic hydrochloric acid-carmine as a stain for chromosomes in squash preparations. Stain Technol. 38:9-13.
Solereder, H.
  1908.   Systematic anatomy of dicotyledons. Vol. II. Monochlamydeae. (Transl. by L. A. Boodle and R. E. Fritsch), Oxford. vi+645-1182.
Srivastava, P. S.
  1971.   In vitro induction of triploid roots and shoots from mature endosperm of *Jatropha panduraefolia*. Z. Pflanzenphysiol. 66:93-96.
Standley, P., and J. A. Steyermark.
  1949.   Flora of Guatemala. Fieldiana Bot. 24(6):126-130.
Stebbins, G. L.
  1950.   Variation and evolution in plants. Columbia University Press, New York. 643pp.
  1957.   Self-fertilization and population variability in the higher plants. Amer. Naturalist 91:338-354.

1958.   Longevity, habitat and release of genetic variability in the higher plants. Cold Spring Harbor Symposia, Quant. Biol. 23:365-378.

1970.   Adaptive radiation in angiosperms. I. Pollination mechanisms. Annual Rev. Ecol. Syst. 1:307-326.

1971.   Chromosomal evolution in higher plants. Addison-Wesley Publ. Co., Reading, Mass. 216pp.

1974.   Flowering plants—evolution above the species level. The Belknap Press, Cambridge, Mass. 399pp.

Tomlinson, P. B., and A. M. Gill.
1973.   Growth habits of tropical trees; some guiding principles, pp. 129-143. *In* B. J. Meggers, E. S. Ayensu, and W. D. Duckworth (eds.), Tropical Africa and South America: a comparative review. Smithsonian Institution Press, City of Washington.

Troll, W.
1935.   Vergleichende Morphologie der Loheren Pflanzen. Bd. 1, Leiferung 2, pp. 101-107. Gebruder Borntraeger, Berlin.

1964.   Die Infloreszenzen. Bd. 1. Gustav Fischer, Stuttgart.

1969.   Die Infloreszenzen. Bd. 2, Teil 1. Gustav Fischer, Stuttgart.

Untawale, A. G., and P. K. Mukherjee.
1969.   Structure and development of glands in *Jatropha gossypifolia* L. J. Indian Bot. Soc. 48:359-362.

Urbatsch, L. E., J. D. Bacon, R. L. Hartman, M. C. Johnston, T. J. Watson, Jr., and G. L. Webster.
1975.   Chromosomal numbers for North American Euphorbiaceae. Amer. J. Bot. 65: 494-500.

Weberling, F.
1965.   Typology of inflorescences. J. Linn. Soc. Bot. 59:215-221.

Webster, G. L.
1956.   A monographic study of the West Indian species of *Phyllanthus* [introduction]. J. Arnold Arb. 37:217-268.

1975.   Conspectus of a new classification of the Euphorbiaceae. Taxon 24:593-601.

Webster, G. L. and D. Burch.
1968.   Flora of Panama; Part VI. Family 97. Euphorbiaceae. Ann. Missouri Bot. Gard. 54:211-350.

Webster, G. L. and L. J. Poveda.
1978.   A phytogeographically significant new species of *Jatropha* (Euphorbiaceae) from Costa Rica. Brittonia 30:265-270.

Webster, G. L. and E. A. Rupert.
1973.   Phylogenetic significance of pollen nuclear number in the Euphorbiaceae. Evolution 27:524-531.

Wilbur, R. L.
1954.   A synopsis of *Jatropha*, subsection *Eucurcas,* with the description of two new species from Mexico. J. Elisha Mitchell Sci. Soc. 70:92-101.

PLATES

PLATE I: GROWTH HABIT

Section *Peltatae* (Pax) Dehgan & Webster

Fig. A.   *J. multifida* L. (B67.282–Jamaica).

Fig. B.   *J. podagrica* (B73.128–India).

Section *Polymorphae* Pax

Fig. C.   *J. macrorhiza* Benth., growth habit (photographed in Arizona, 9 miles northeast of Aricava Junction).

Fig. D.   *J . macrorhiza* Benth., subterranean tuberous portion.

Fig. E.   *J. macrantha* Muell. Arg. (courtesy of Mr. W. G. Roberts–photographed in Peru).

Fig. F.   *J. integerrima* Jacq. (B67.280–Cuba).

PLATE II: GROWTH HABIT

Section *Tuberosae* Pax

Fig. A.   *J. woodii* O. Ktze. (B75.267—Natal).

Fig. B.   *J. capensis* (L. f.) Sond.; note the existence of a caudex and the true subterranean tuber (B67.045—South Africa).

Section *Collenucia* (Chiov.) Chiov.

Fig. C.   *J. paradoxa* (Chiov.) Chiov. (B75.023—Somali Republic).

Fig. D.   *J. marginata* Chiov. (B75.039—Somali Republic).

Section *Jatropha*
Subsection *Pubescentes* Pax ex Dehgan & Webster

Fig. E.   *J. velutina* Pax & Hoffm. (B75.048—Kenya).

PLATE III: GROWTH HABIT

Fig. A. *J. cordata* (Ortega) Muell. Arg. (photographed 4 miles west of Concordia, Mexico).

Fig. B. *J. curcas* L. (B68.028–Ghana).

Fig. C. *J. cinerea* (Ortega) Muell. Arg. (photographed 4.5 miles north of Mulege, March, 1974).

Fig. D. *J. canescens* (Benth.) Muell. Arg.; shows the much branched growth habit characteristic of this species. (photographed 2 miles north of Huatabampito, west coast of Mexico, December 1974).

Fig. E. *J. cinerea* (Ortega) Muell. Arg.; note the above ground roots. (photographed at Bahía Kino, Sonora, Mexico, December, 1974).

PLATE IV: GROWTH HABIT

Fig. A. *J. dioica* Sessé (photographed in the Chihuahuan Desert, December, 1974).

Fig. B. *J. cardiophylla* (Torr.) Muell. Arg. (photographed 9.6 miles south of Sandario Road, Arizona, June, 1974).

Fig. C. *J. dioica* Sessé (B74.226–Chihuahuan Desert). Note the rhizomatous growth habit (arrow).

Fig. D. *J. cuneata* Wiggins & Rollins (photographed in Bahía San Carlos, Mexico, December, 1974), a more typical form of this species in both the west coast of Mexico and Baja California.

Fig. E. *J. cuneata* Wiggins & Rollins (photographed 9 miles north and 4 miles east of La Paz near sand dunes, Baja California, March, 1974); note the extremely deep roots in relation to size of the plant. This particular population is the only one known with a chromosome number of more than 2 x = 44.

Fig. F. *J. cuneata* Wiggins & Rollins (photographed about 6 miles north of San Ignacio in Baja California, March, 1974); a rather unusual form with bright yellow stems.

PLATE V: CLEARED LEAVES

Section *Polymorphae* Pax

Fig. A. *J. hernandiifolia* Vent. (B67.281–Jamaica), X 0.5.

Fig. B. *J. divaricata* Swartz (from U.C.D. Herbarium; *Webster et al. 8387*–Jamaica), ca. X 0.5.

Fig. C. *J. macrorhiza* Benth. (B74.076–Arizona), ca. X 0.3.

Fig. D. *J. tupifolia* Griseb. (from U.C.D. Herbarium; *Webster et al. 198*–Cuba), ca. X 0.5.

Fig. E. *J. integerrima* Jacq. (B67.280–Cuba), ca. X 0.5.

Fig. F. *J. hastata* Jacq.; this leaf has been cleared to show the distinction between *J. integerrima* Jacq. and *J. hastata* Jacq. as is evident from the shape of the leaf and venation pattern (from U.C.D. Herbarium; collected in Coral Gables Bot. Gard., Univ. of Miami; *Webster et al.*), ca. X 0.5.

Fig. G. *J. angustifolia* Griseb. (from U.C.D. Herbarium; *Webster et al. 4673*–Cuba), ca. X 2.0.

PLATE VI: CLEARED LEAVES

Section *Tuberosae* Pax

Fig. A.   *J. unicostata* Balf. f. (B67.471–Socotra), X 1.0.

Fig. B.   *J. gallabatensis* Schweinf. (B74.161–Somali Republic), X 1.0.

Fig. C.   *J. capensis* (L.f.) Sond. (B67.045–South Africa), X 1.0.

Section *Peltatae* (Pax) Dehgan & Webster

Fig. D.   *J. augustii* Pax & K. Hoffm. (B74.056–Peru), part.

Fig. E.   *J. multifida* L. (B63.016–Ghana) part.

Fig. F.   *J. cathartica* Terán & Berland. (B67.471–Texas), part.

Section *Collenucia* (Chiov.) Chiov.

Fig. G.   *J. paradoxa* (Chiov.) Chiov. (B75.023–Somali Republic), X 1.0.

Section *Jatropha*

Fig. H.   *J. brockmanii* Hutchinson (B75.070–Kenya), X 1.0.

Fig. I.   *J. velutina* Pax & Hoffm. (B75.048– Kenya), X 1.0.

Fig. J.   *J. gossypiifolia* L. (B74.016–India), X 0.5.

PLATE VII: CLEARED LEAVES

Section *Curcas* (Adans.) Griseb.

Fig. A. *J. curcas* L. (B68.286—Senegal), ca. X 0.5.

Fig. B. *J. malacophylla* Standley (B71.072—Mexico), ca. X 0.5.

Fig. C. *J. pseudocurcas* Muell. Arg. (from U.C.D. Herbarium, *Webster & Adams 11746E*—Chiapas, Mexico), part.

Section *Platyphyllae* Dehgan & Webster

Fig. D. *J. ciliata* Sessé (B75.051—Mexico), ca. X 0.5.

Fig. E. *J. platyphylla* Muell. Arg. (B69.300—Mexico), ca. X 0.25.

Fig. F. *J. moranii* Dehgan & Webster (B75.052—Baja California), ca. X 1.0.

PLATE VIII: CLEARED LEAVES

Section *Loureira* (Cav.) Muell. Arg. ex Pax

Fig. A. *J. gaumeri* Greenm. (from U.C.D. Herbarium); doubtfully placed here since the specimen is incomplete, ca. × 0.5.

Fig. B. *J. canescens* (Benth.) Muell. Arg. (B74.195—Mexico), × 1.0.

Fig. C. *J. cordata* (Ortega) Muell. Arg. (B74.204—Sonora, Mexico), × 1.0.

Fig. D. *J. cinerea* (Ortega) Muell. Arg. (B74.035—Baja California), × 1.0.

Fig. E. *J. standleyi* Steyerm. (B75.047—Mexico), × 0.5.

Fig. F. *J. ortegae* Standley ? (B74.232—Mexico), × 1.0.

PLATE IX: CLEARED LEAVES

Section *Loureira* (Cav.) Muell. Arg. ex Pax
Subsection *Neopauciflorae* Dehgan & Webster

Fig. A. *J. fremontioides* Standley (B74.163—Mexico), ca. X 2.0.

Fig. B. *J. neopauciflora* Pax (from U.C.D. Herbarium, *Webster 17319*—Mexico), ca. X 2.0.

Fig. C. *J. sympetala* Blake & Standley (from U.C.D. Herbarium, *Webster 17292*—Mexico), ca. X 2.0.

Section *Mozinna* (Ortega) Pax

Fig. D. *J. cuneata* Wiggins & Rollins; (upper) normal leaf; (lower) short shoot leaf. (B74.025—Baja California), ca. X 2.0.

Fig. E. *J. dioica* Sessé; (left) normal leaf; (right) short shoot leaf. (B67.067—Mexico), X 1.0. X 1.0.

Fig. F. *J. dioica* Sessé (upper) normal; (lower) short shoot leaf. (B67.279—Mexico), X 2.0.

Fig. G. *J. cardiophylla* (Torr.) Muell. Arg. (B74.072—Arizona), ca. X2.0.

Figs. E and F are both included here to point out the possibility of their taxonomic distinctness.

PLATE X: STIPULES

Fig. A.   *J. augustii* Pax (detail of Fig. B), ca. X 4.5.

Fig. B.   *J. augustii* Pax (B74.056–Peru), ca. X 2.0.

Fig. C.   *J. podagrica* Hook. (B73.129–Bogor, India), ca. X 3.0.

Fig. D.   *J. curcas* L. (B73.128–India), X 3.0.

Fig. E.   *J. integerrima* Jacq. (B67.280–Cuba), ca. X 3.0.

Fig. F.   *J. macrorhiza* Benth. (B74.076–Arizona), ca. X 3.0.

Fig. G.   *J. ferox* Pax, spinose stipules (B75.268–Somali Republic).

Fig. H.   *J. dioica* Sessé (B74.223–Mexico), X 5.0.

Fig. I.   *J. cordata* Muell. Arg. (B74.006–Mexico), ca. X 5.0.

Fig. J.   *J. gossypiifolia* L., petiolar and stipular glands (B74.016–India).

Fig. K.   *J. gossypiifolia* L., stipitate glands of the petiole (B74.016–India).

Fig. L.   *J. gossypiifolia* L., long-section of foliar gland (B74.016–India).

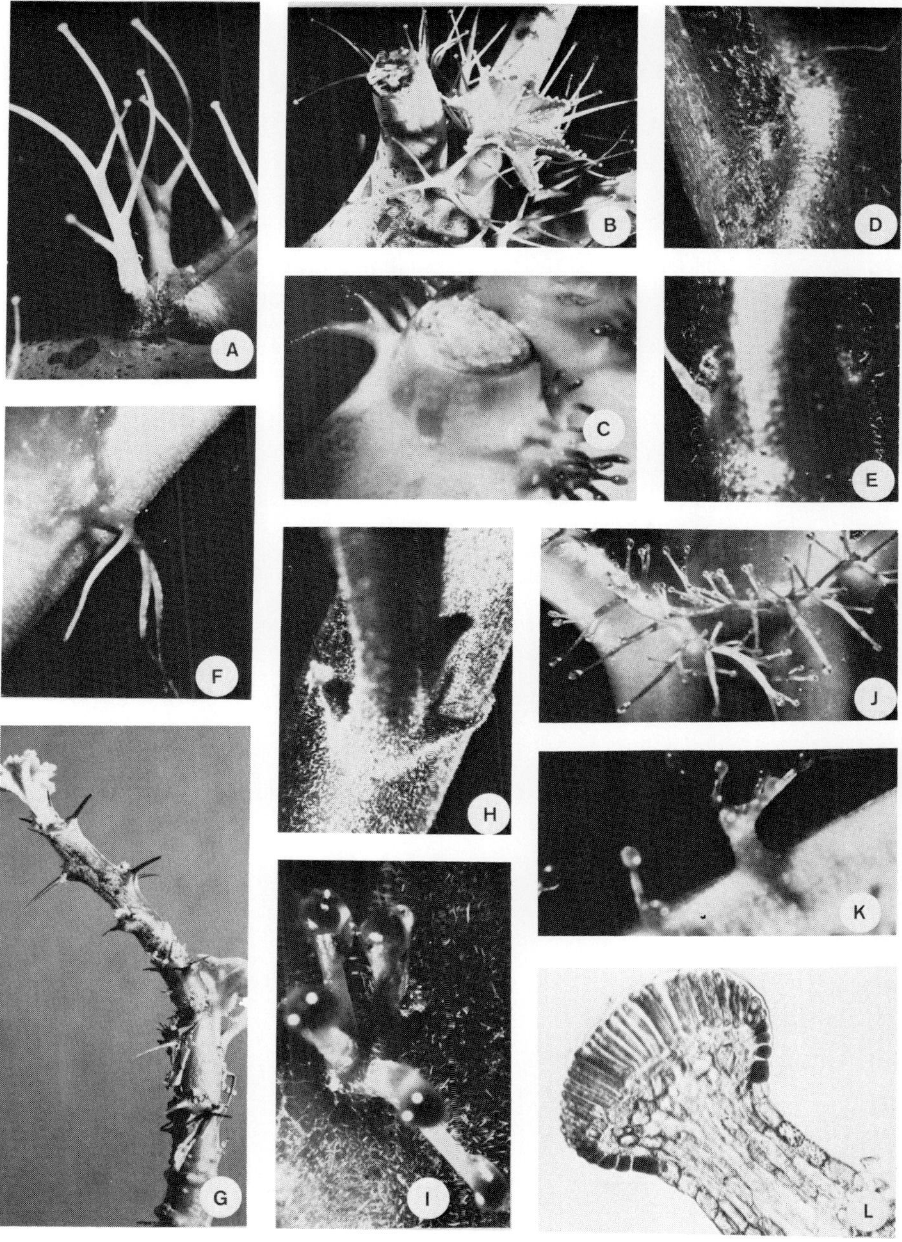

PLATE XI: INFLORESCENCE

Section Peltatae (Pax) Dehgan & Webster

Fig. A. *J. podagrica* Hook. (B73.129—India), detail of a portion of the inflorescence, indicating the dichasial character and early opening of the pistillate flowers.

Fig. B. *J. multifida* L. (B63.016—Ghana), pistillate flowers have abscised and only staminate flowers remain.

Fig. C. *J. augustii* Pax (B74.056—Peru), pseudo-terminal position of the inflorescence is indicated as the growth of shoot meristem is somewhat delayed but continuous.

Fig. D. *J. augustii* Pax (B74.056—Peru), detail of the inflorescence, only pistillate flowers are open.

Section Tuberosae Pax

Fig. E. *J. unicostata* Balf. f. (B67.471—Socotra), production of inflorescence has resulted in growth of lateral buds (arrows).

Fig. F. *J. capensis* (L.f.) Sond. (B67.045—South Africa), terminal inflorescence.

Fig. G. *J. glauca* Vahl (B75.050—Aden), inflorescence.

Fig. H. *J. glauca* Vahl (B75.050—Aden), portion of the inflorescence showing central position of the pistillate flower in dichasia.

PLATE XII: INFLORESCENCE

Section *Curcas* (Adans.) Griseb.

Fig. A. *J. curcas* L. (B74.039–Guyana), main florescence on right, coflorescence on left.

Fig. B. *J. curcas* L. (B74.039–Guyana), detail of the main florescence; note the single pistillate flower at the center.

Fig. C. *J. curcas* L. (B74.039–Guyana), portion of the inflorescence showing the replacement of pistillate flower with a staminate one.

Fig. D. *J. mcvaughii* Dehgan & Webster (B74.206–Mexico), pistillate inflorescence.

Fig. E. *J. mcvaughii* Dehgan & Webster (B74.206–Mexico), staminate inflorescence.

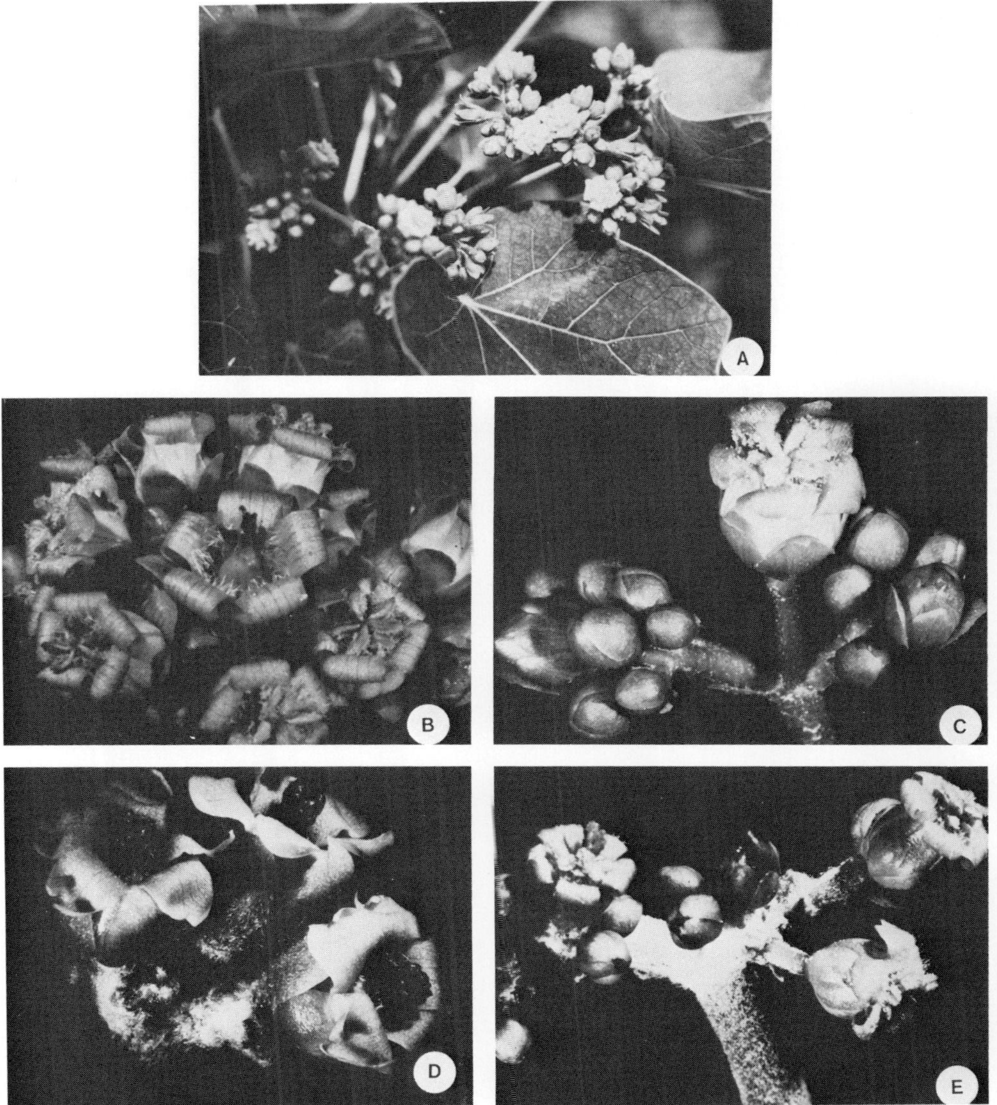

PLATE XIII: INFLORESCENCE

Section *Polymorphae* Pax

Fig. A.   *J. integerrima* Jacq. (B67.280–Cuba), two distinct florescences are shown; because of the close proximity, one resembles a coflorescence.

Fig. B.   *J. integerrima* Jacq. (B67.280–Cuba), detail of the inflorescence, only pistillate flowers are open.

Fig. C.   *J. macrorhiza* Benth. (B74.075–Arizona), secondary inflorescences which have appeared following senescence of the terminal inflorescence (arrow).

Fig. D.   *J. macrorhiza* Benth. (B74.075–Arizona), enlargement of the central infloresence.

Fig. E.   *J. hernandiifolia* Vent. (B67.281–Jamaica), symmetrical inflorescence with a pistillate flower at the center and each dichasium also terminating in a pistillate flower.

Fig. F.   *J. hernandiifolia* Vent. (B67.281–Jamaica), same as Fig. B., showing further development of the inflorescence.

PLATE XIV: INFLORESCENCE

Section *Jatropha*

Fig. A. *J. gossypiifolia* L. (B74.040–Guyana), terminal main florescence and coflorescence
are nearly indistinguishable; the portion on the right, however, has fewer flowers
and is somewhat shorter and could therefore safely be considered as the coflorescence.

Fig. B. *J. gossypiifolia* L. (B74.040–Guyana), detail of a portion of the florescence

Section *Collenucia* (Chiov.) Chiov.

Fig. C. *J. marginata* Chiov. (B74.162–Somali Republic).

Fig. D. *J. paradoxa* (Chiov.) Chiov. (B75.023–Somali Republic).

Note that in both species the coflorescence has been reduced to a single pistillate flower.

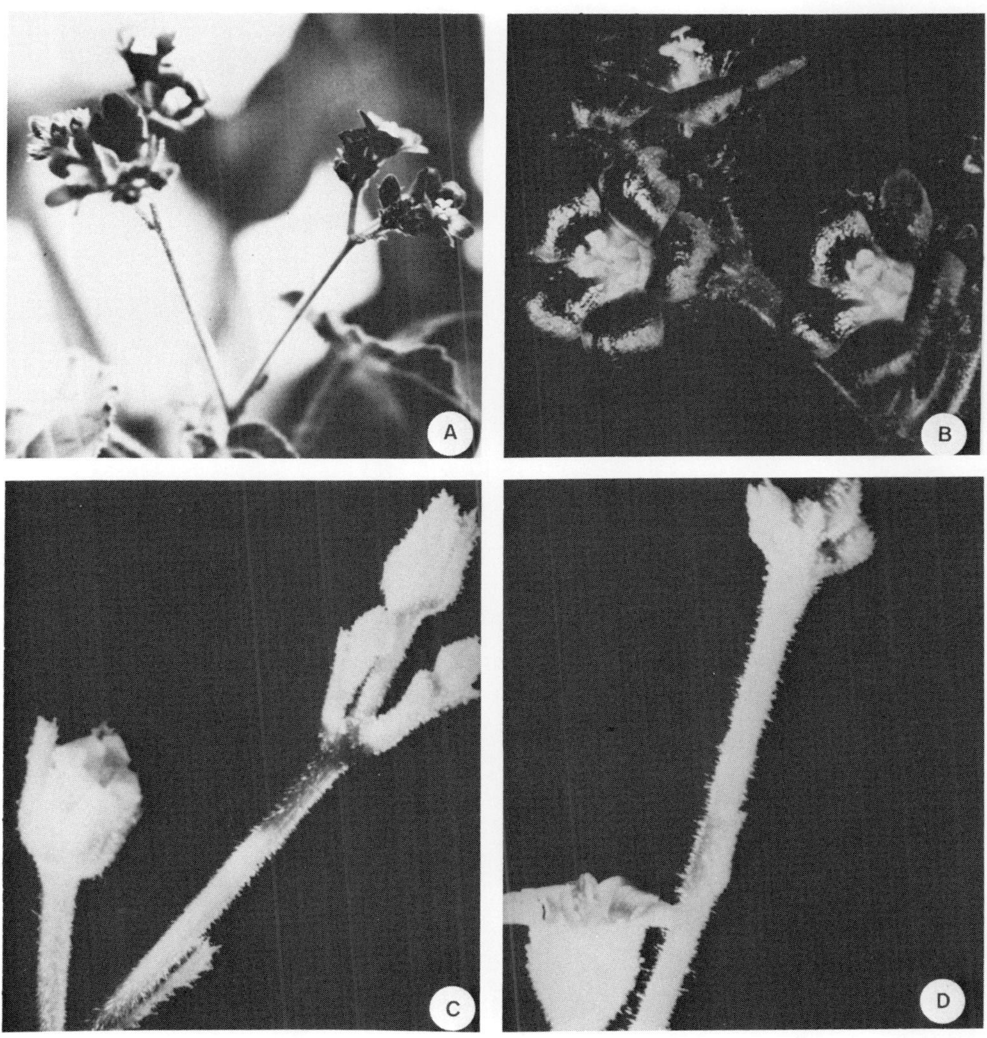

PLATE XV: INFLORESCENCE

Section *Platyphyllae* Dehgan & Webster

Fig. A. *J. platyphylla* Muell. Arg. (B69.300–Mexico), pistillate inflorescence.

Fig. B. *J. platyphylla* Muell. Arg. (B74.207–Mexico), part of the staminate inflorescence.

Section *Loureira* (Cav.) Muell. Arg. ex Pax

Fig. C. *J. cinerea* (Ortega) Muell. Arg. (B74.028–Magdalena Bay, Baja California), terminal inflorescence.

Fig. D. *J. canescens* (Benth.) Muell. Arg. (B74.038–Magdalena Island, Baja California), inflorescence lateral in the leaf axil.

Fig. E. *J. cordata* (Ortega) Muell. Arg. (B74.204–Sonora, Mexico); the inflorescence is actually terminal, but lateral shoots have grown since the initiation of the inflorescence. Sepals of the central pistillate flower are shown by arrow.

Fig. F. *J. canescens* (Benth.) Muell. Arg. (B74.004–Navajoa, Mexico); note the lateral position of the inflorescence in the leaf axil.

PLATE XVI: INFLORESCENCE

Section *Mozinna* (Ortega) Pax

Fig. A. *J. dioica* Sessé (B67.279–Mexico), female inflorescence consisting of only two flowers.

Fig. B. *J. dioica* Sessé (B71.070–Mexico), male inflorescence with the entire branch suggestive of a compound dichasium.

Fig. C. *J. dioica* Sessé (B71.070–Mexico), detail of a male inflorescence in Fig. B; note the number of male flowers as compared to Fig. A.

PLATE XVII: FLOWERS

Section *Curcas* (Adans.) Griseb.

Fig. A. *J. curcas* L. (B74.039–Guyana), pistillate flower, X 3.0.

Fig. B. *J. curcas* L. (B74.039–Guyana), detail of Fig. A., X 3.5.

Fig. C. *J. curcas* L. (B74.039–Guyana), staminate flower, X 3.0.

Fig. D. *J. curcas* L. (B74.039–Guyana), detail of Fig. C., X 5.0.

Fig. E. *J. mcvaughii* Dehgan & Webster (B74.206–Mexico), pistillate flower, X 4.0.

Fig. F. *J. mcvaughii* Dehgan & Webster (B74.206–Mexico), staminate flower, X 5.5

Section *Platyphyllae* Dehgan & Webster

Fig. G. *J. platyphylla* Muell. Arg. (B74.194–Mexico), pistillate flower, X 1.5

Fig. H. *J. platyphylla* Muell. Arg. (B74.194–Mexico), detail of Fig. G., X 2.5.

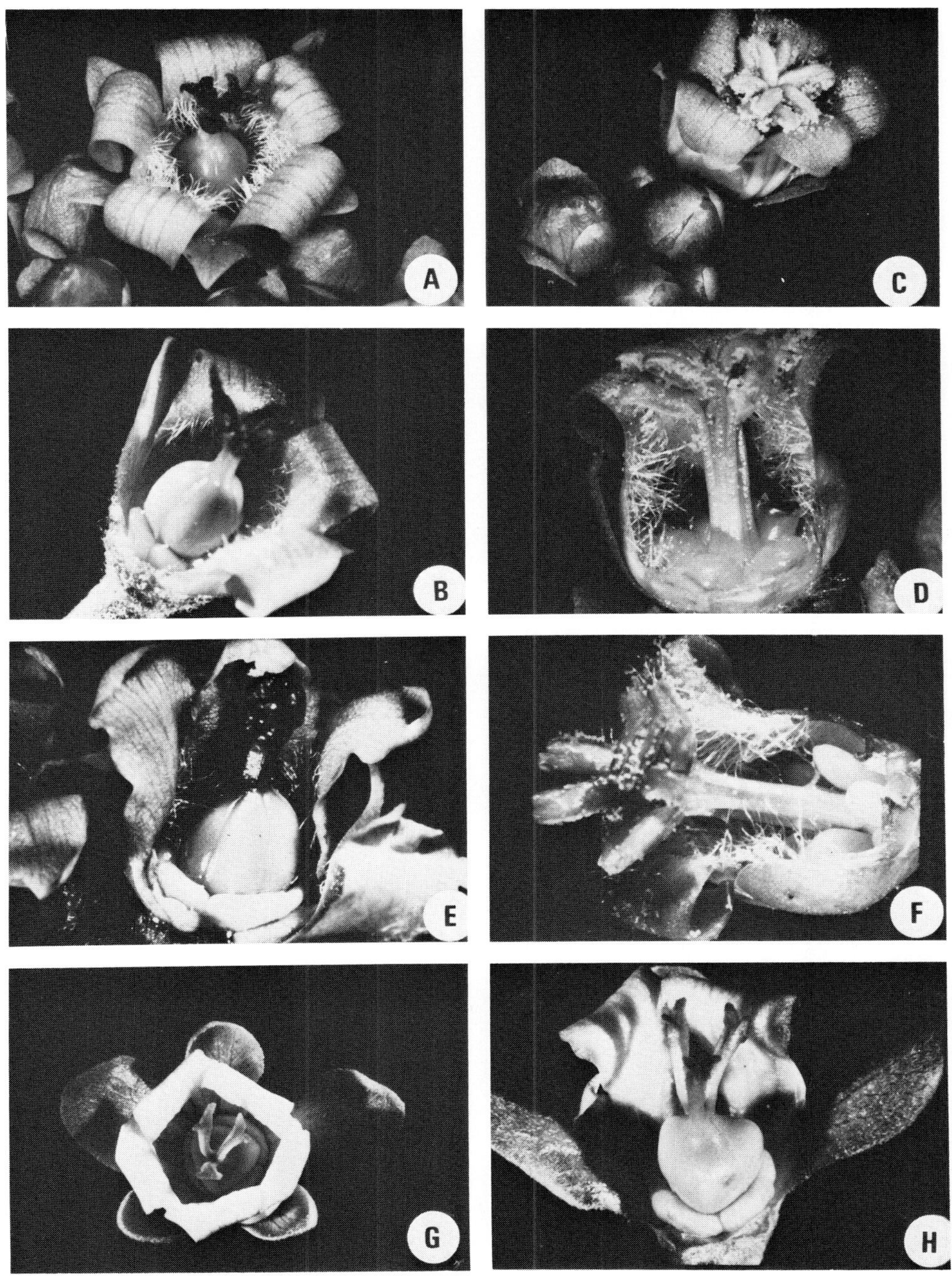

PLATE XVIII: FLOWERS

Section *Polymorphae* Pax

Fig. A.   *J. macrorhiza* Benth. (B75.075—Arizona), pistillate flower, X 1.0.

Fig. B.   *J. macrorhiza* Benth. (B75.075—Arizona), detail of Fig. A., X 3.0.

Fig. C.   *J. macrorhiza* Benth. (B75.075—Arizona), staminate flower, X 2.0.

Fig. D.   *J. macrorhiza* Benth. (B75.075—Arizona), detail of Fig. C., X 4.0.

Fig. E.   *J. integerrima* Jacq. (B67.280—Cuba), pistillate flower, X 1.0.

Fig. F.   *J. integerrima* Jacq. (B67.280—Cuba), detail of Fig. E., X 6.0.

Fig. G.   *J. integerrima* Jacq. (B67.280—Cuba), staminate flower, X 2.0.

Fig. H.   *J. integerrima* Jacq. (B67.280—Cuba), detail of Fig. G., X 4.0.

Fig. I.   *J. hernandiifolia* Vent. (B67.281—Jamaica), pistillate flower, X 3.0.

Fig. J.   *J. hernandiifolia* Vent. (B67.281—Jamaica), staminate flower, X 4.5.

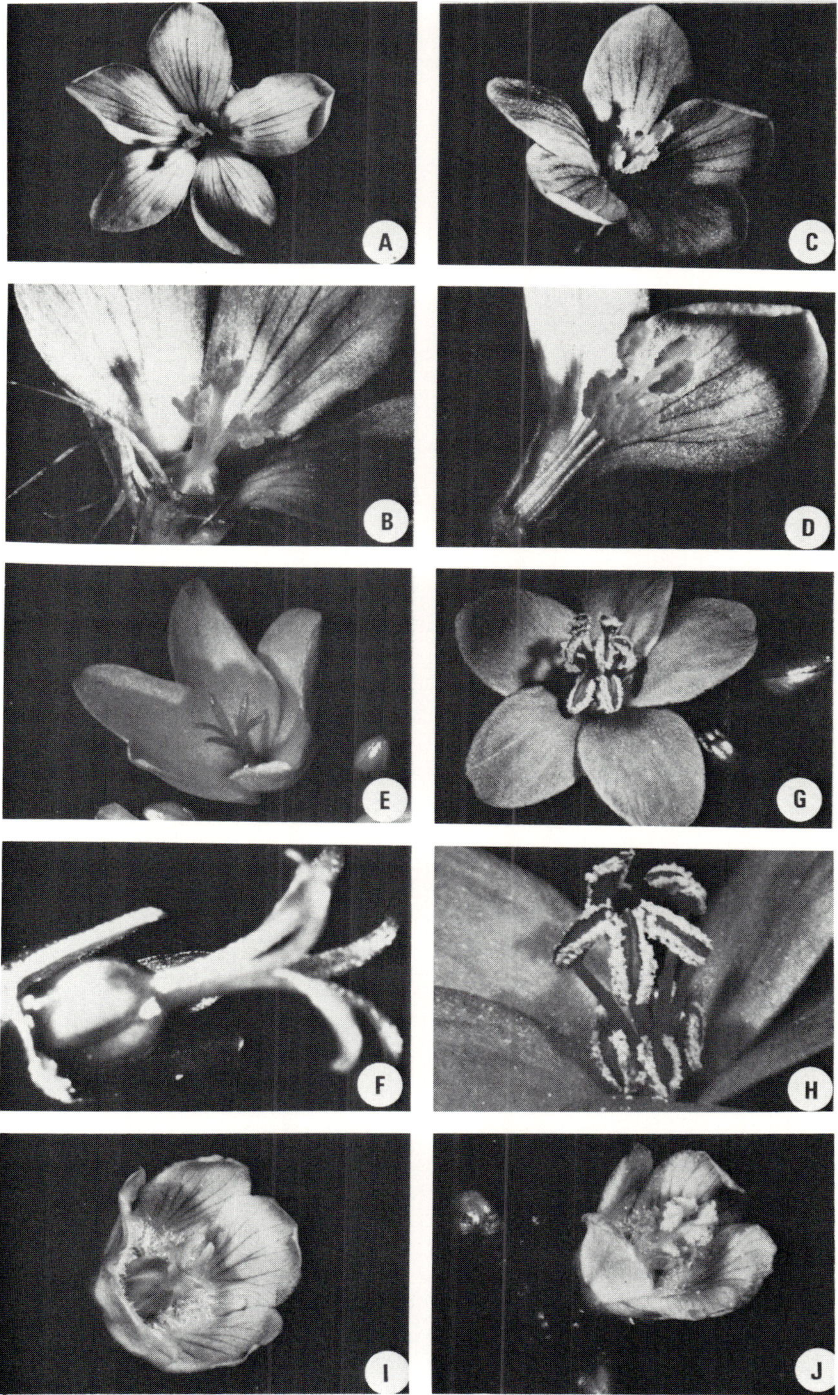

PLATE XIX: FLOWERS

Section *Peltatae* (Pax) Dehgan & Webster

Fig. A. *J. cathartica* Terán & Berland. (B75.057–Texas), pistillate flower, $\times$ 2.0.

Fig. B. *J. cathartica* Terán & Berland. (B75.057–Texas), detail of Fig. A., $\times$ 4.5.

Fig. C. *J. cathartica* Terán & Berland. (B75.057–Texas), staminate flower, $\times$ 3.0.

Fig. D. *J. cathartica* Terán & Berland. (B75.057–Texas), detail of Fig. C., $\times$ 4.5.

Fig. E. *J. podagrica* Hook. (B73.129–India), pistillate flowers, $\times$ 1.5.

Fig. F. *J. podagrica* Hook. (B73.129–India), staminate flower, $\times$ 1.5.

Fig. G. *J. multifida* L. (B63.016–Ghana), pistillate flower, $\times$ 1.0.

Fig. H. *J. multifida* L. (B63.016–Ghana), staminate flower, $\times$ 3.0.

Fig. I. *J. augustii* Pax & Hoffm. (B74.056–Peru), pistillate flower, $\times$ 2.0.

Fig. J. *J. augustii* Pax & Hoffm. (B74.056–Peru), detail of Fig. I., $\times$ 6.0.

Fig. K. *J. augustii* Pax & Hoffm. (B74.056–Peru), staminate flower, $\times$ 2.0.

Fig. L. *J. augustii* Pax & Hoffm. (B74.056–Peru), detail of Fig. K., $\times$ 6.0.

PLATE XX: FLOWERS

Section *Tuberosae* Pax
Subsection *Capenses* Dehgan & Webster

Fig. A. *J. capensis* (L.f.) Sond. (B67.045—South Africa), pistillate flower, X 4.0.

Fig. B. *J. capensis* (L.f.) Sond. (B67.045—South Africa), detail of Fig. A., X 6.0.

Fig. C. *J. capensis* (L.f.) Sond. (B67.045—South Africa), flower staminate, X 4.0.

Fig. D. *J. capensis* (L.f.) Sond. (B67.045—South Africa), detail of Fig. C., X 6.0.

Fig. E. *J. glauca* Vahl (B75.050—Aden), pistillate flower, X 3.0.

Fig. F. *J. glauca* Vahl (B75.050—Aden), detail of Fig. E., X 5.0.

Fig. G. *J. glauca* Vahl (B75.050—Aden), staminate flower, X 4.0.

Fig. H. *J. glauca* Vahl (B75.050—Aden), detail of Fig. G., X 5.0.

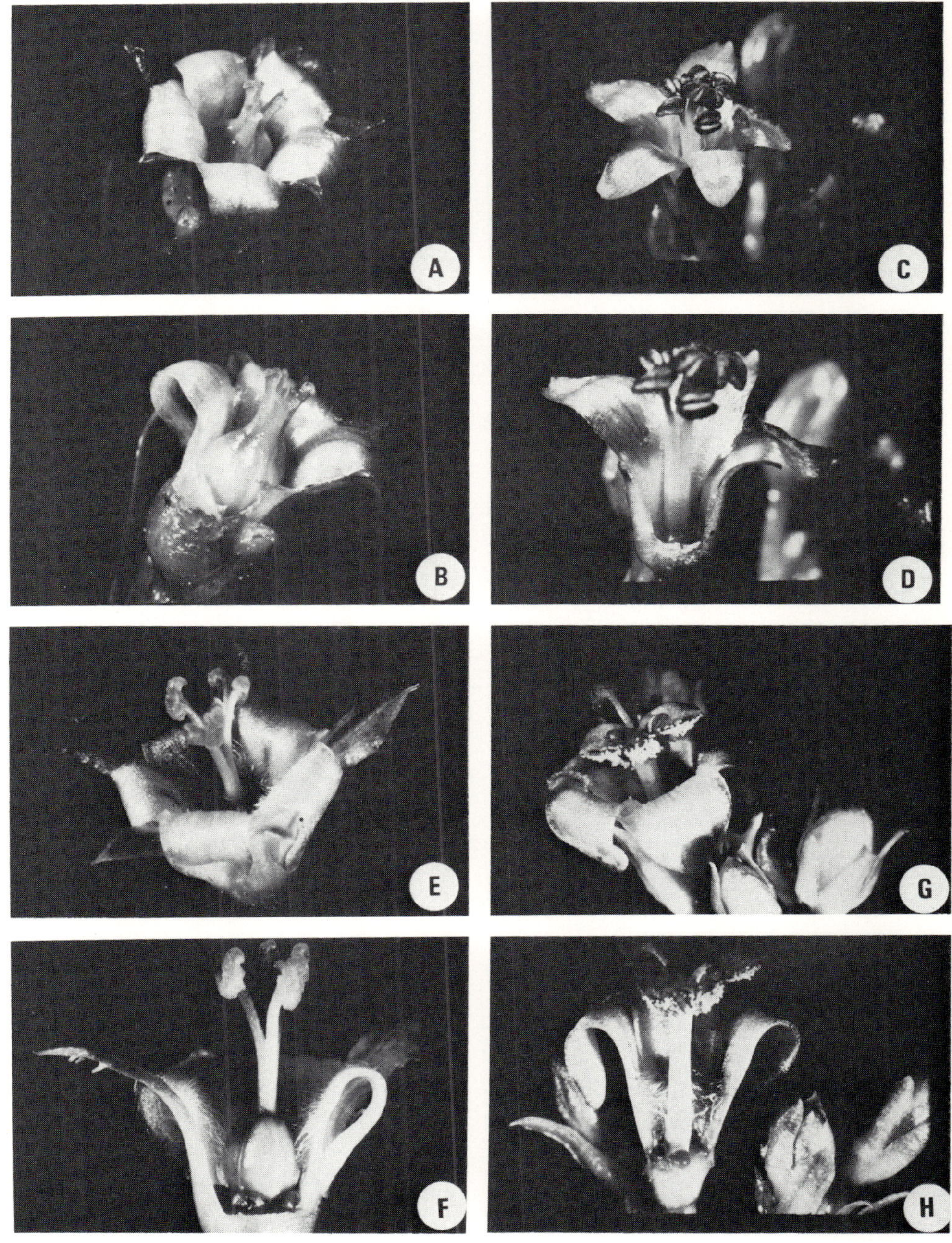

PLATE XXI: FLOWERS

Section *Tuberosae* Pax
Subsection *Tuberosae* Dehgan & Webster

Fig. A. *J. unicostata* Balf. f. (B67.471–Socotra), pistillate flower, X 1.3.

Fig. B. *J. unicostata* Balf. f. (B67.471–Socotra), detail of Fig. A., X 3.0.

Fig. C. *J. unicostata* Balf. f. (B67.471–Socotra), staminate flower, X 1.3.

Fig. D. *J. unicostata* Balf. f. (B67.471–Socotra), detail of Fig. B., X 4.0.

Fig. E. *J. gallabatensis* Schweinf. (B74.161–Somali Republic), pistillate flower, X 3.0.

Fig. F. *J. gallabatensis* Schweinf. (B74.161–Somali Republic), detail of Fig. E., X 4.0.

Fig. G. *J. gallabatensis* Schweinf. (B74.161–Somali Republic), staminate flower, X 4.0.

Fig. H. *J. gallabatensis* Schweinf. (B74.161–Somali Republic), detail of Fig. G., X 6.0.

PLATE XXII: FLOWERS

Section *Jatropha*
Subsection *Adenophorae* Pax ex Dehgan & Webster

Fig. A. *J. gossypiifolia* L. (B74.016–India), an aberrant pistillate flower with aborted stamens (arrows), X 3.0.

Fig. B. *J. gossypiifolia* L. (B74.016–India), staminate flower X 3.0.

Fig. C. *J. gossypiifolia* L. (B74.016–India), X 3.0.

Section *Spinosae* Pax

Fig. D. *J. fissispina* Pax (B76.002–Tanzania), pistillate flower, X 3.0.

Fig. E. *J. fissispina* Pax (B76.002–Tanzania), detail of Fig. D., X 4.0.

Fig. F. *J. fissispina* Pax (B76.002–Tanzania), staminate flower, X 3.0.

Fig. G. *J. fissispina* Pax (B76.002–Tanzania), detail of Fig. F., X 3.0.

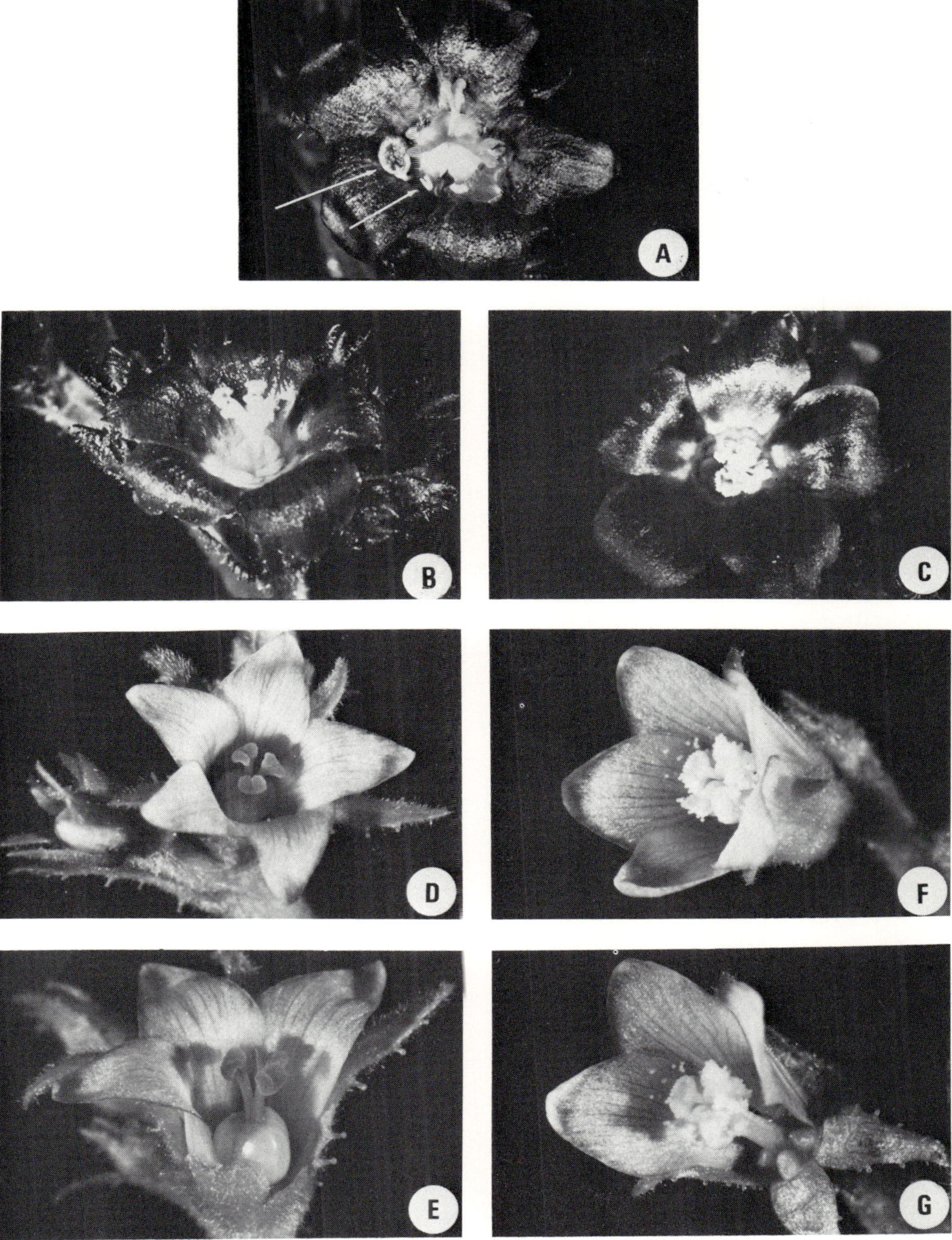

PLATE XXIII: FLOWERS

Section *Jatropha*
Subsection *Pubescentes* ex Dehgan & Webster

Fig. A.   *J. velutina* Pax & Hoffm. (B75.048–Kenya), pistillate flower, X 4.0.

Fig. B.   *J. velutina* Pax & Hoffm. (B75.048–Kenya), staminate flower, X 4.0.

Fig. C.   *J. velutina* Pax & Hoffm. (B75.048–Kenya), detail of Fig. B., X 8.0.

Fig. D.   *J. brockmanii* Hutchinson (B75.070–Kenya), pistillate flower, X4.0.

Fig. E.   *J. brockmanii* Hutchinson (B75.070–Kenya), detail of Fig. D., X 8.0.

Fig. F.   *J. brockmanii* Hutchinson (B75.070–Kenya), staminate flower, X 6.5.

Fig. G.   *J. brockmanii* Hutchinson (B75.070–Kenya), detail of Fig. F., X 11.0.

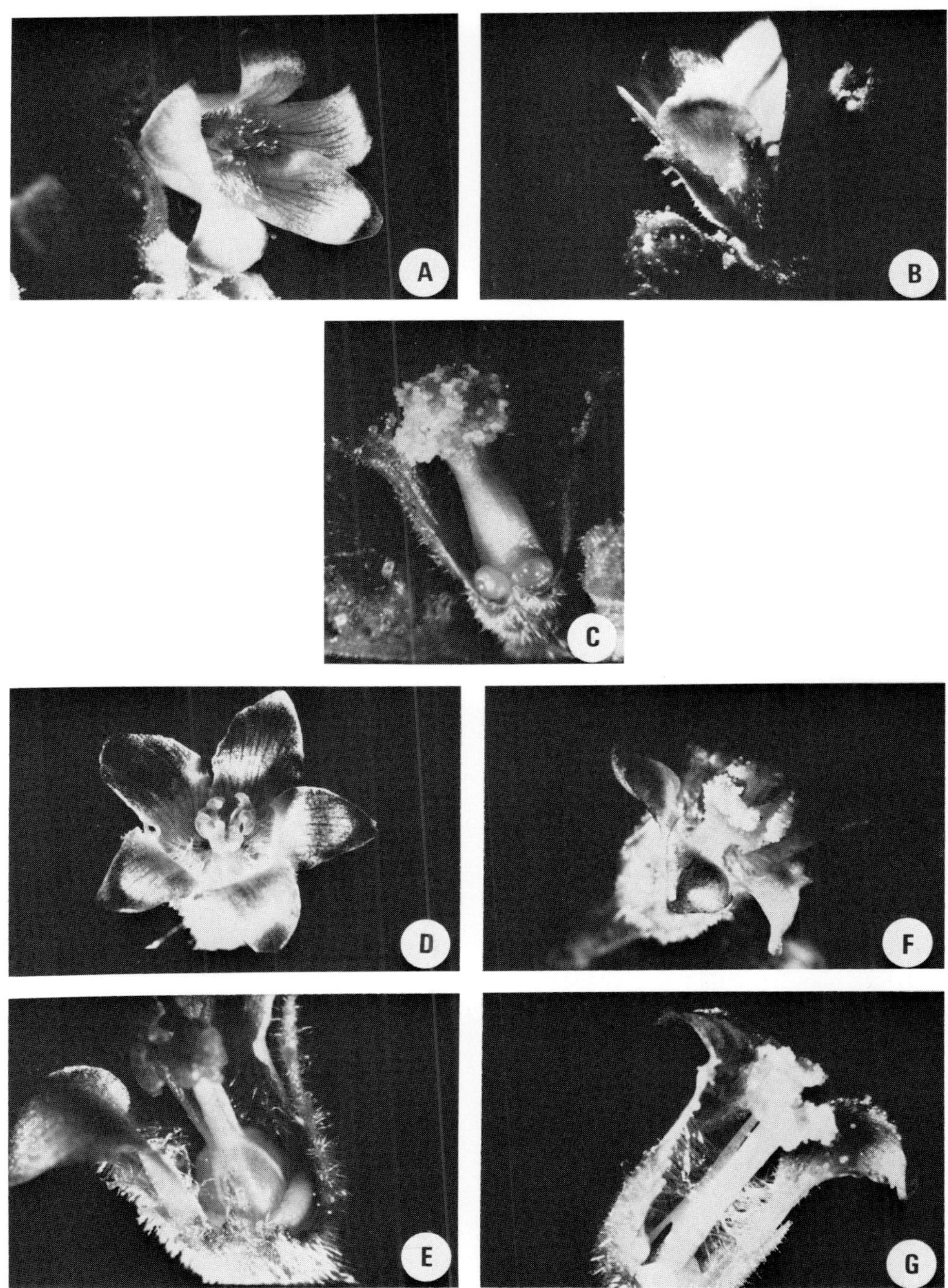

PLATE XXIV: FLOWERS

Section *Collenucia* (Chiov.) Chiov.

Fig. A. *J. paradoxa* (Chiov.) Chiov. (B75.023–Somali Republic), pistillate flower, X 5.0.

Fig. B. *J. paradoxa* (Chiov.) Chiov. (B75.023–Somali Republic), detail of Fig. A., X 7.0.

Fig. C. *J. paradoxa* (Chiov.) Chiov. (B75.023–Somali Republic), staminate flower, X 9.0.

Fig. D. *J. paradoxa* (Chiov.) Chiov. (B75.023–Somali Republic), detail of Fig. C., X 7.0.

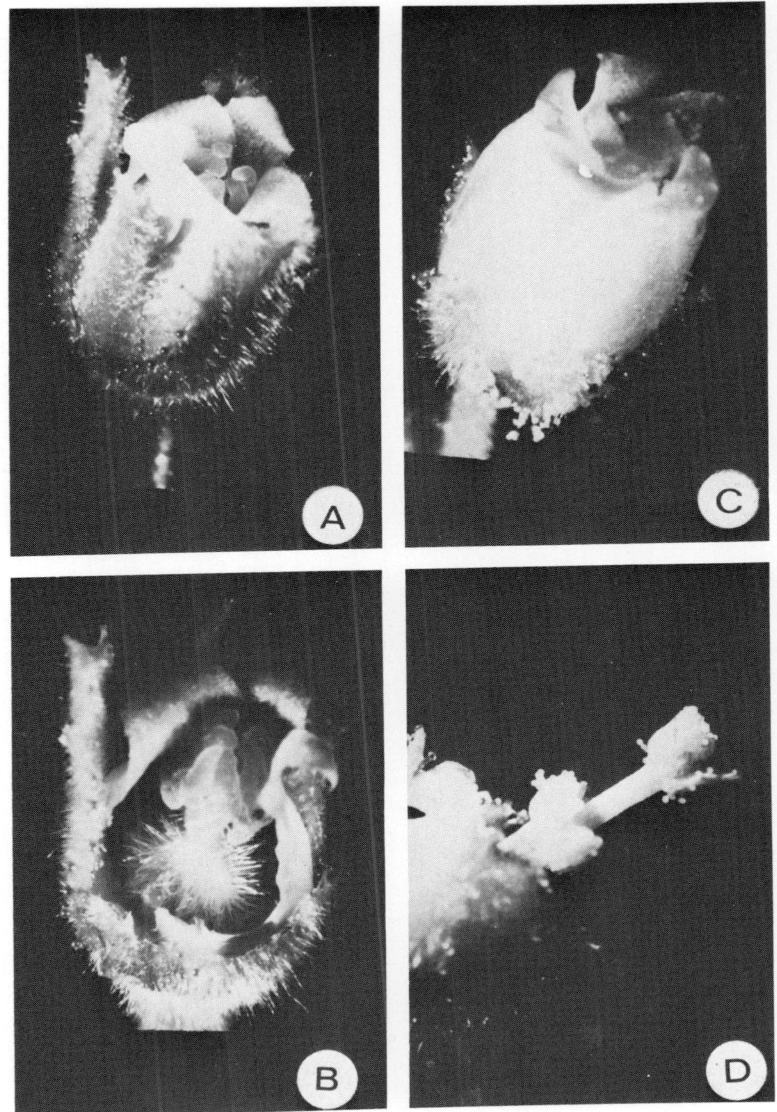

PLATE XXV: FLOWERS

Section *Loureira* (Cav.) Muell. Arg. ex Pax
Subsection *Loureira* (Cav.) Muell. Arg.

Fig. A.　*J. cordata* (Ortega) Muell. Arg. (B74.204–Sonora, Mexico), pistillate flower, X 1.5.

Fig. B.　*J. cordata* (Ortega) Muell. Arg. (B74.204–Sonora, Mexico), detail of Fig. A., X 5.0.

Fig. C.　*J. cordata* (Ortega) Muell. Arg. (B74.204–Sonora, Mexico), staminate flower, X 3.0.

Fig. D.　*J. cordata* (Ortega) Muell. Arg. (B74.204–Sonora, Mexico), detail of Fig. C., X 8.0.

Subsection *Canescentes* Pax ex Dehgan & Webster

Fig. E.　*J. canescens* (Benth.) Muell. Arg. (B74.195–Mexico), pistillate flower, X 3.5

Fig. F.　*J. canescens* (Benth.) Muell. Arg. (B74.195–Mexico), detail of Fig. A., X 4.0.

Fig. G.　*J. canescens* (Benth.) Muell. Arg. (B74.195–Mexico), staminate flower, X 5.0.

Fig. H.　*J. canescens* (Benth.) Muell. Arg. (B74.195–Mexico), detail of Fig. C., X 9.0.

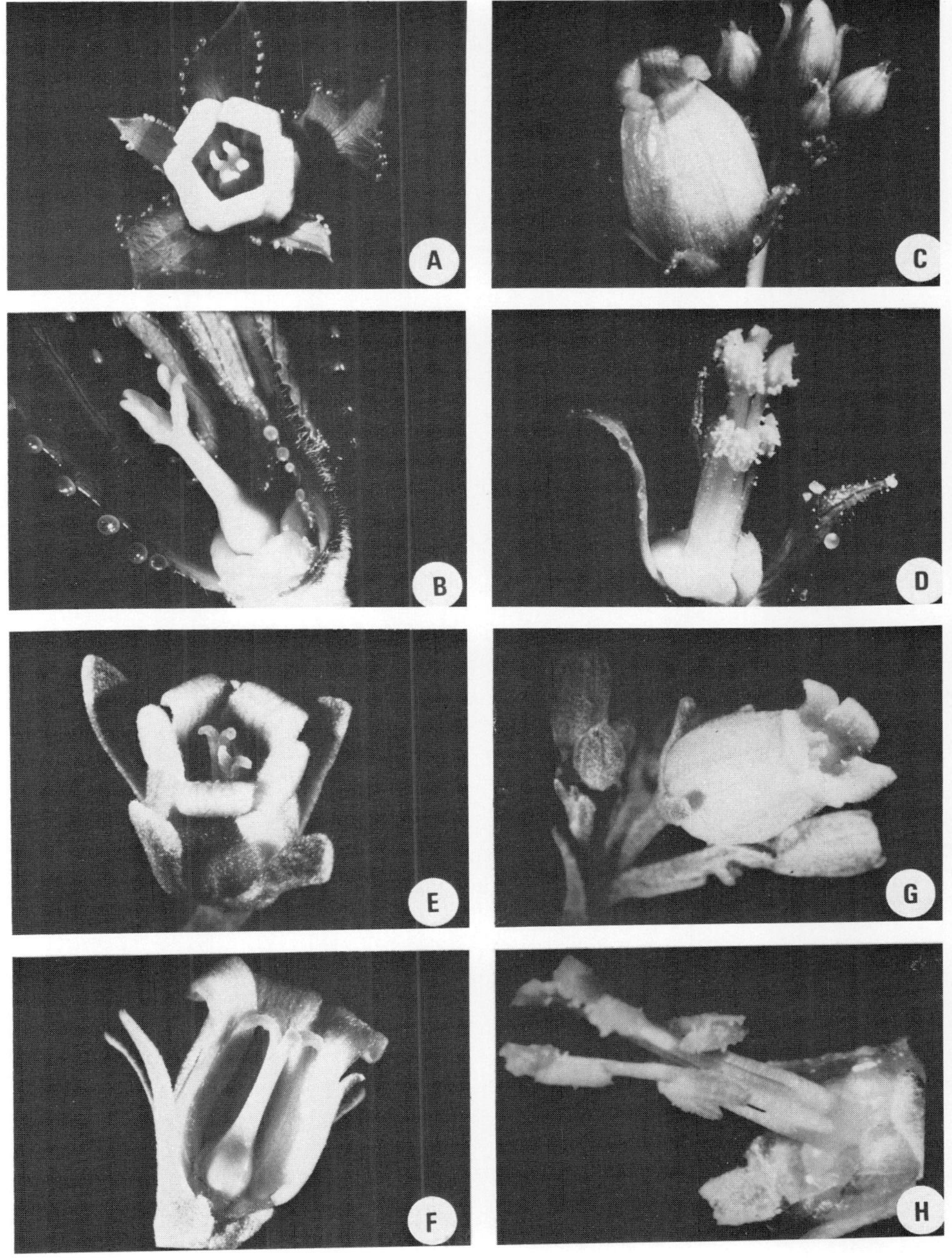

PLATE XXVI: FLOWERS

Section *Mozinna* (Ortega) Pax

Fig. A. *J. dioica* Sessé (B74.231–Mexico), pistillate flower, X 2.5.

Fig. B. *J. dioica* Sessé (B74.231–Mexico), detail of Fig. A., X 4.5.

Fig. C. *J. dioica* Sessé (B74.231–Mexico), staminate flower, X 3.5.

Fig. D. *J. dioica* Sessé (B74.231–Mexico), detail of Fig. C., X 5.0.

Fig. E. *J. cardiophylla* (Torr.) Muell. Arg. (B74.072–Arizona), detail of pistillate flower, X 3.0.

Fig. F. *J. cardiophylla* (Torr.) Muell. Arg. (B74.071–Arizona), staminate flower, X 6.0.

Fig. G. *J. cuneata* Wiggins & Rollins (B74.025–Baja California), pistillate flower, X 3.0.

Fig. H. *J. cuneata* Wiggins & Rollins (B74.200–Mexico), staminate flower, X 3.5.

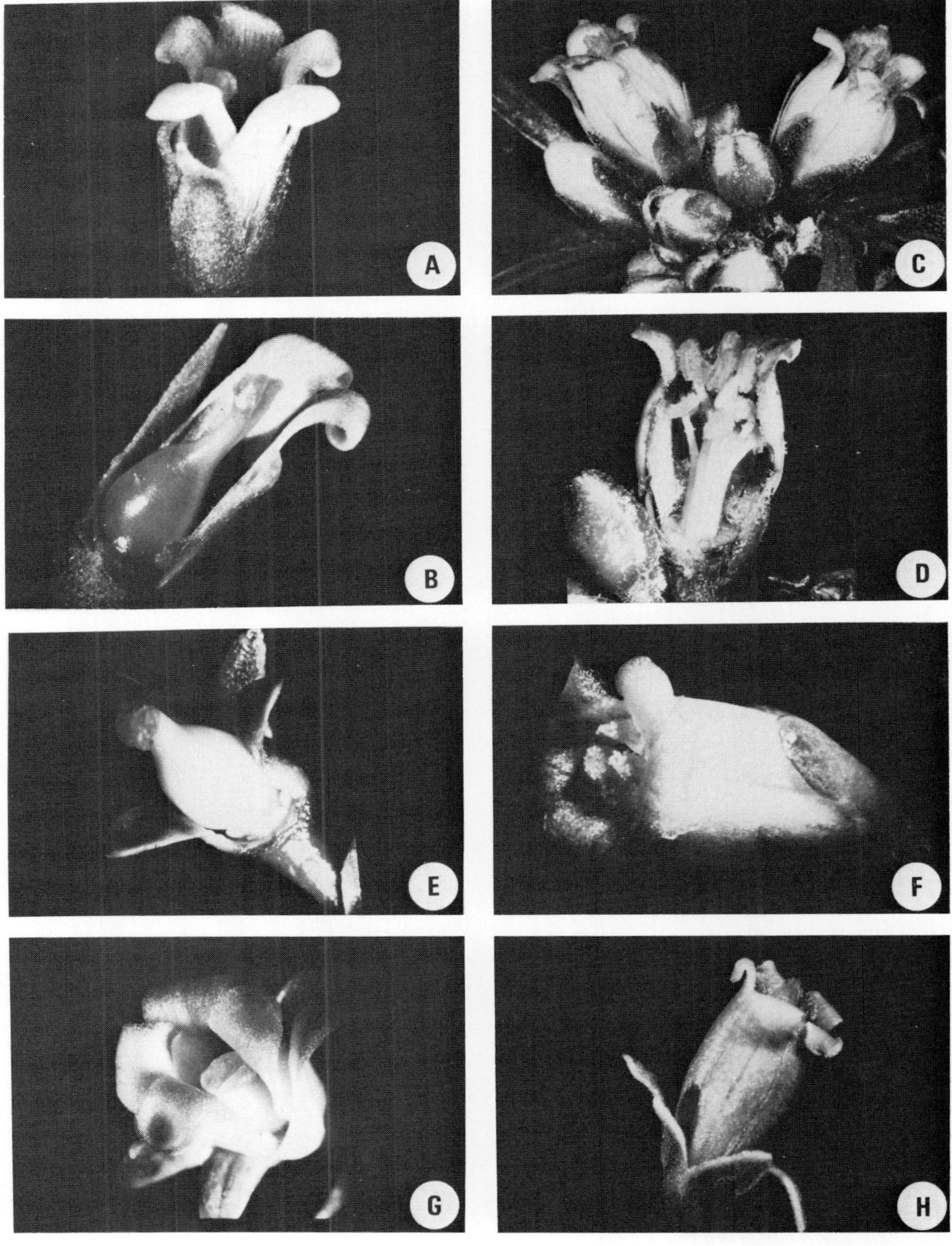

PLATE XXVII: POLLEN GRAINS

Fig. A. *J. gallabatensis* Schweinf. (B74.161—Somali Republic), whole grain, X 1020.

Fig. B. *J. gallabatensis* Schweinf. (B74.161—Somali Republic), detail of the striated exinous knobs, X 3800.

Fig. C. *J. gossypiifolia* L. (B74.016—India), X 10,600.

Fig. D. *J. integerrima* Jacq. (B67.280—Cuba), non-striated exinous knobs, X 2700.

Fig. E. *J. integerrima* Jacq. (B67.280—Cuba), striated exinous knobs in the pollen of a different plant, X 6075.

Fig. F. *J. capensis* (L.f.) Sond. (B67.410—South Africa), X 7700.

Fig. G. *J. macrorhiza* Benth. (B74.075—Arizona), X 4320.

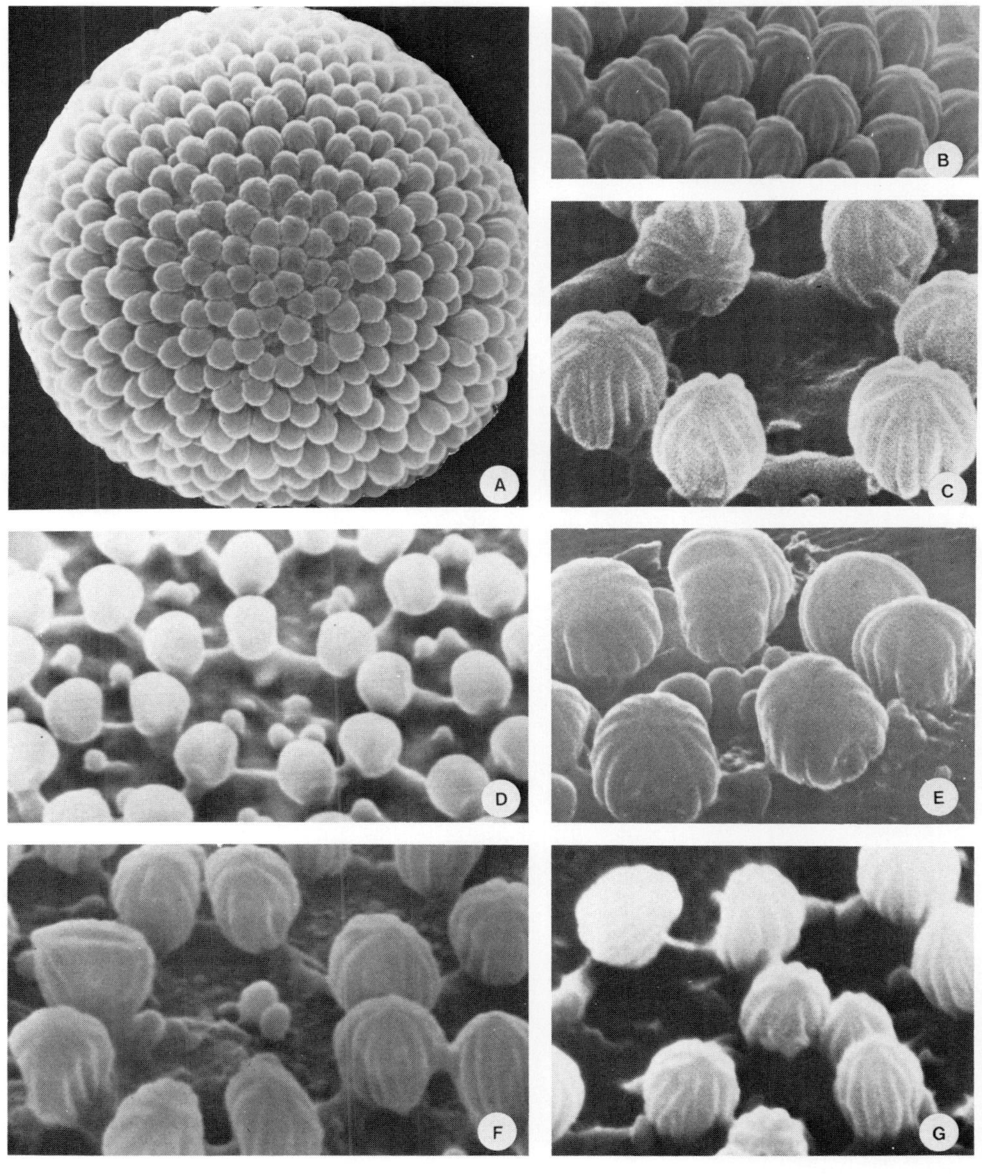

PLATE XXVIII: POLLEN GRAINS

Fig. A.   *J. cathartica* Terán & Berland. (B67.524—Texas), whole grain, X 400 (courtesy of S. P. Lynch).

Fig. B.   *J. cathartica* Terán & Berland. (B67.524—Texas), detail of smooth exinous knobs, X 2900.

Fig. C.   *J. cathartica* Terán & Berland. (B67.524—Texas), detail of the arrangement of the exinous knobs, X 1000.

Fig. D.   *J. podagrica* Hook. (B59.047—Ghana), detail of a non-striated exinous knobs, X 1550.

Fig. E.   *J. augustii* Pax & Hoffm. (B74.056—Peru), also with smooth exinous knobs, X 950.

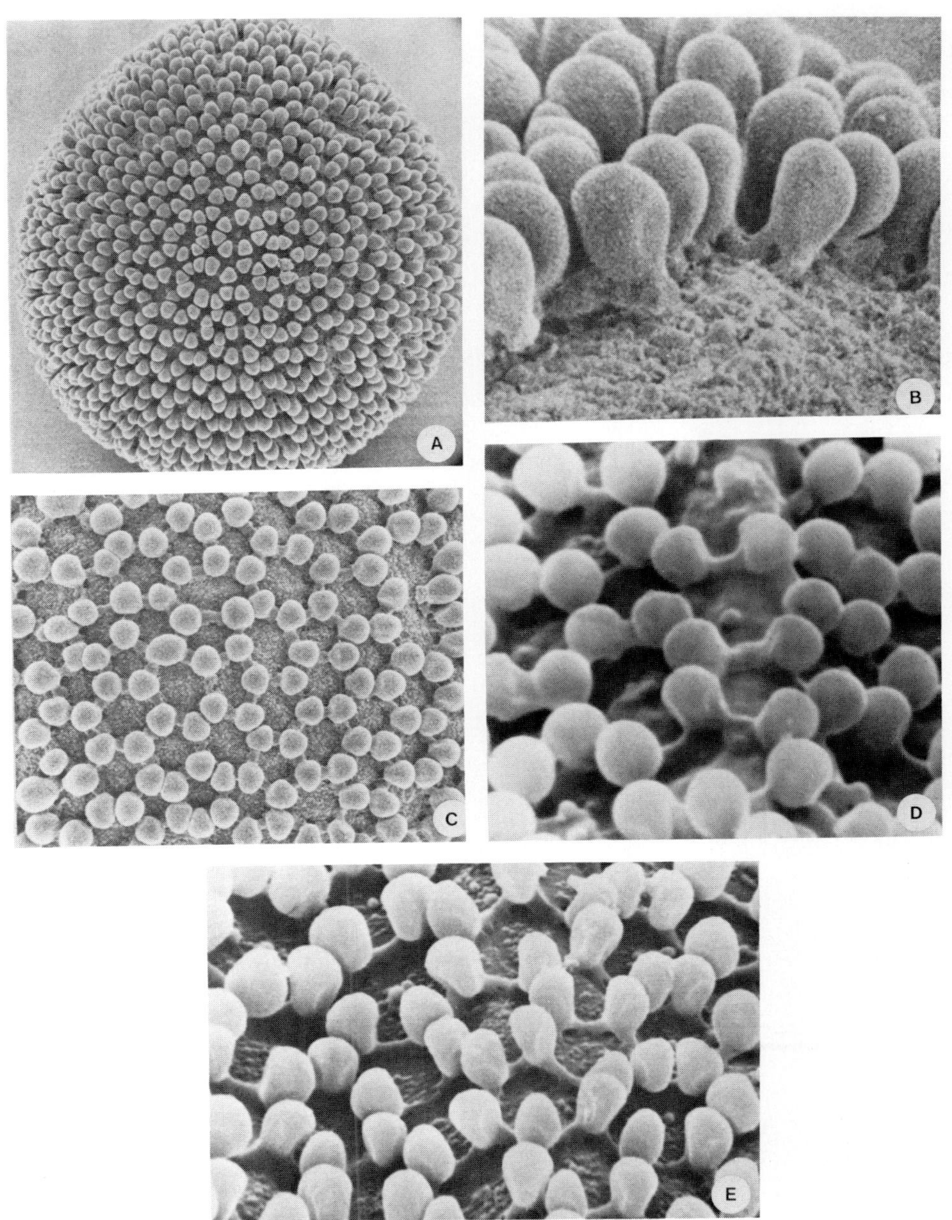

PLATE XXIX: POLLEN GRAINS

Fig. A. *J. dioica* Sessé (B71.070—Mexico), whole grain, X 625 (courtesy of S. P. Lynch).

Fig. B. *J. dioica* Sessé (B71.070—Mexico), detail of the arrangement of exinous knobs, X 3100.

Fig. C. *J. canescens* (Benth.) Muell. Arg. (B71.069—Mexico), X 2500.

Fig. D. *J. curcas* L. (B68.286—Senegal), X 2750.

Fig. E. *J. moranii* Dehgan & Webster (B75.052—Baja California), X 3500.

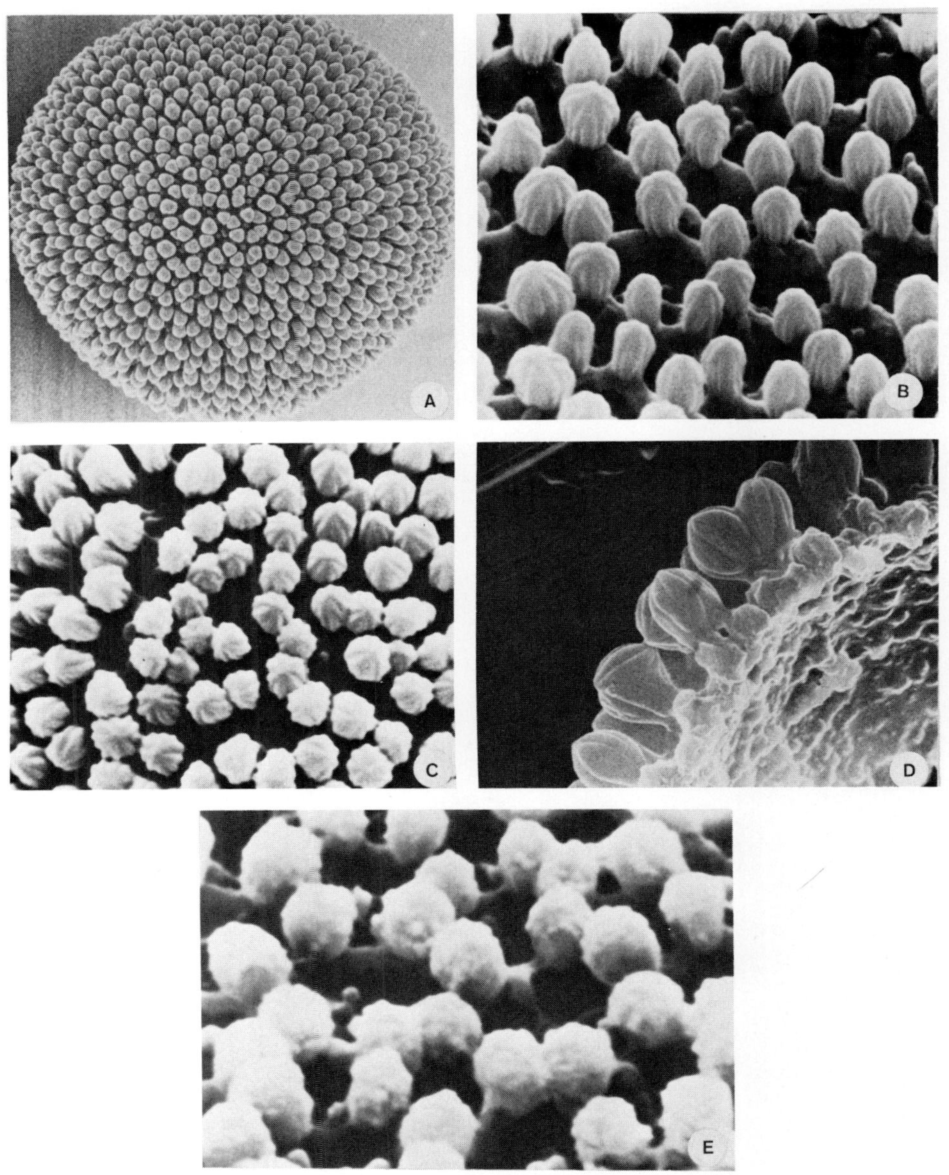

PLATE XXX: FRUITS

Fig. A. *J. gossypiifolia* L. (B74.166–South America), X 1.3.

Fig. B. *J. brockmanii* Hutchinson (B75.048–Kenya), X 1.5.

Fig. C. *J. integerrima* Jacq. (B67.282)–Jamaica), X 1.3.

Fig. D. *J. multifida* L. (B67.280–Cuba), X 0.5.

Fig. E. *J. macrorhiza* Benth. (B74.074–Arizona), X 1.0.

Fig. F. *J. unicostata* Balf. f. (B67.471–Socotra), X 1.0.

Fig. G. *J. hernandiifolia* Vent. (B67.281–Jamaica), X 1.5.

Fig. H. *J. curcas* L. (B68.144–Senegal), X 0.5.

Fig. I. *J. dioica* Sessé (B74.223–Mexico), X 1.0.

Fig. J. *J. cardiophylla* (Torr.) Muell. Arg. (B74.072–Arizona) X 1.0; note the lateral position of the fruit.

Fig. K. *J. dioica* Sessé (B67.279–Mexico), X 1.0. Terminal position of the fruit in this photograph and the lateral position in Fig. I. are shown here to indicate the possibility of the distinctiveness of *J. dioica* and *J. spathulata.*

Fig. L. *J. cordata* (Ortega) Muell. Arg. (B74.204–Mexico), X 2.5.

PLATE XXXI: SEEDLINGS

Fig. A. *J. hieronymii* O. Ktze. (B74.068—Argentina).

Fig. B. *J. curcas* L. (B74.041—Sierra Leone).

Fig. C. *J. multifida* L. (B75.029—Angola); note the cryptocotylar germination of the species as the cotyledons have remained inside the seed (arrow).

Fig. D. *J. macrorhiza* Benth. (B74.076—Arizona).

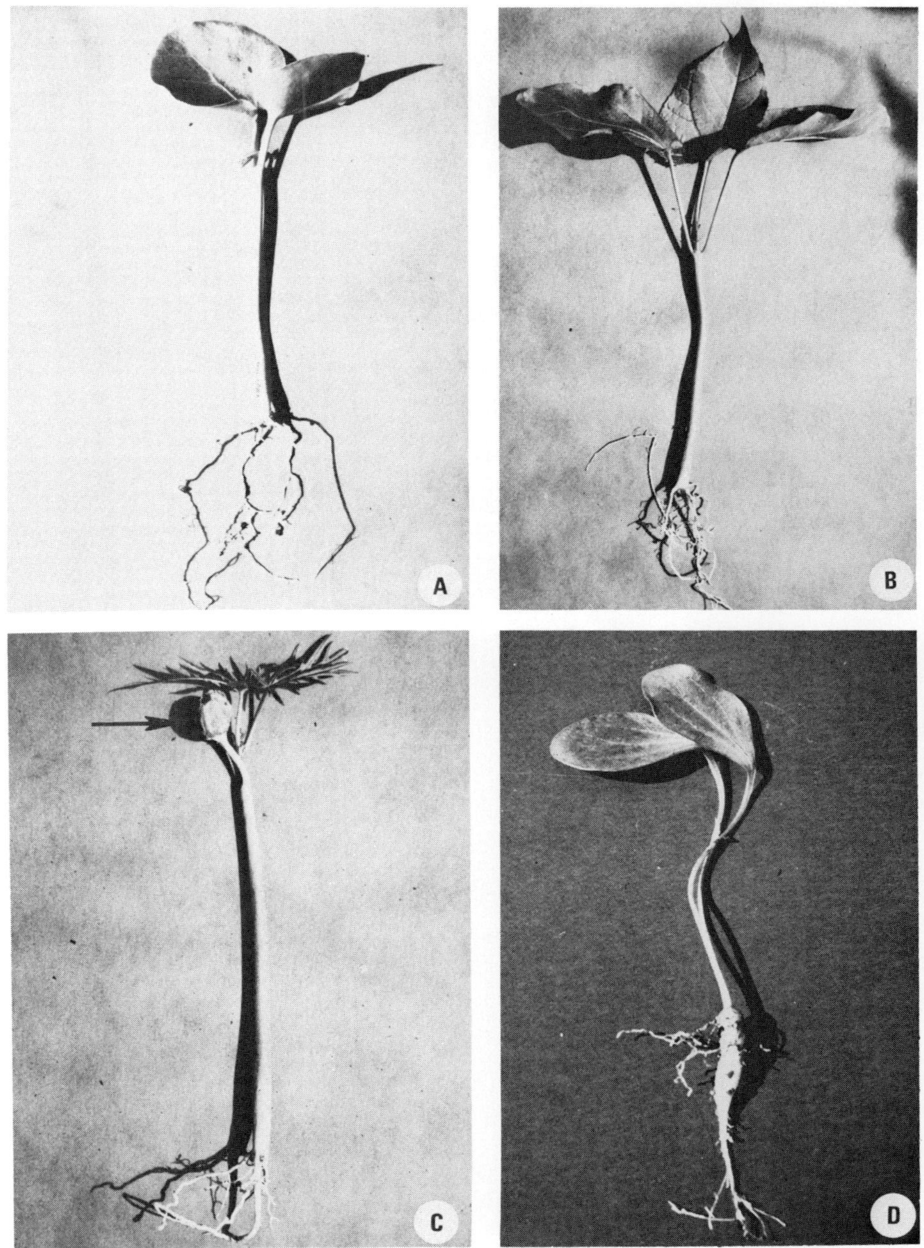

PLATE XXXII: SEEDS

Fig. A.   *J. hieronymii* O. Ktze. (B74.068—Argentina).

Fig. B.   *J. podagrica* Hook. (B74.011—Puerto Rico).

Fig. C.   *J. multifida* L. (B74.157—New Guinea).

Fig. D.   *J. gossypiifolia* L. (B74.040—Guyana).

Fig. E.   *J. macrorhiza* Benth. (B74.076—Arizona).

Fig. F.   *J. integerrima* Jacq. (B67.280—Cuba).

Fig. G.   *J. capensis* (L.f.) Sond. (B67.045—South Africa).

Fig. H.   *J. brockmanii* Hutchinson (B75.070—Kenya).

Fig. I.   *J. hernandiifolia* Vent. (B67.281—Jamaica).

Fig. J.   *J. unicostata* Balf. f. (B67.471—Socotra).

All seeds are magnified approximately X 1.5.

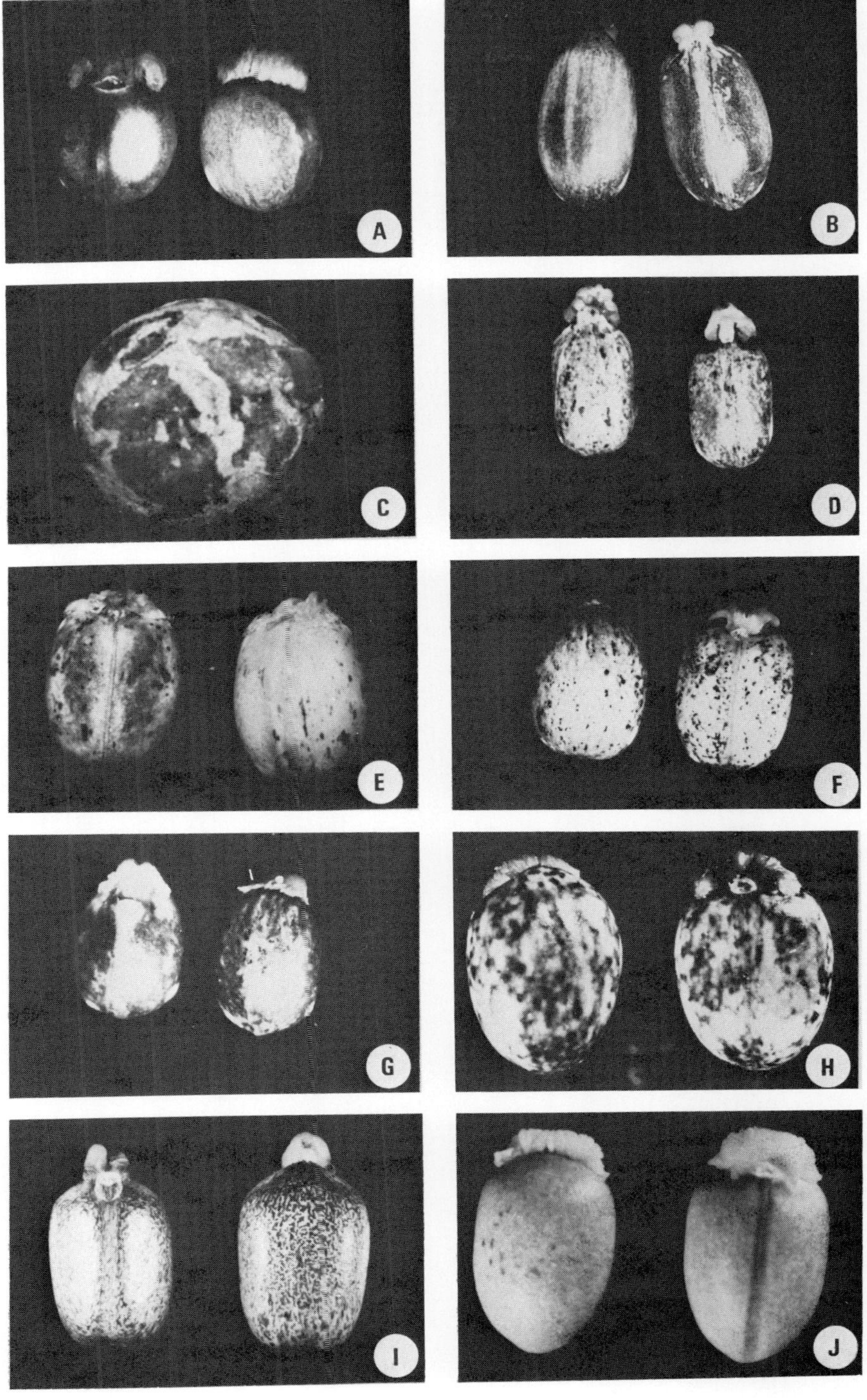

PLATE XXXIII: SEEDS

Fig. A. *J. curcas* L. (B74.012–Puerto Rico), ca. × 1.0.

Fig. B. *J. cardiophylla* (Torr.) Muell. Arg. (B74.072–Arizona), ca. × 1.5.

Fig. C. *J. dioica* Sessé (B74.225–Mexico), ca. × 2.0.

Fig. D. *J. platyphylla* Muell. Arg. (B69.300–Mexico), ca. × 1.5.

Fig. E. *J. cinerea* (Ortega) Muell. Arg. (B74.195–Mexico), ca. × 1.5.

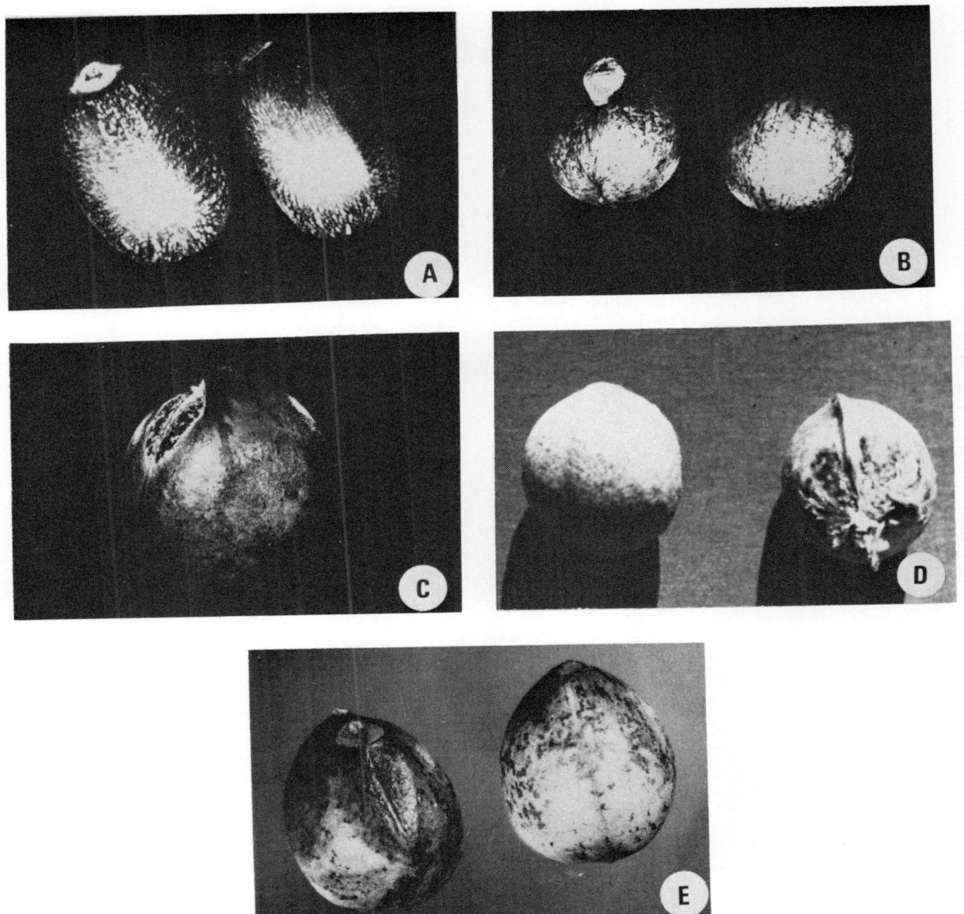